DANIEL KNOP · *Giant Clams*

Dedicated
to all Members
of the
»Clam Family«

Cover illustration: *Tridacna crocea,* raised in captivity by Marine Science Institute,
University of the Philippines (UPMSI)

Daniel Knop

Giant Clams

A Comprehensive Guide to the Identification
and Care of Tridacnid Clams

Dähne Verlag Ettlingen

Knop, Daniel
Giant Clams
A Comprehensive Guide to the Identification and Care of Tridacnid Clams

ISBN 3-921684-23-4

© 1996 Dähne Verlag GmbH, Postfach 250, D-76256 Ettlingen

Translated from the German by Dr. Eva Hert and Dr. Sebastian Holzberg

Photographs without source reference provided by the author.

Cover layout: Bomans Design, Siebeldingen

Layout: Werner Trauthwein

Printed by Kraft Druck GmbH, Ettlingen.

Contents

Preface

The ancient Greeks called the sea "Thalassa". It was curse and blessing at the same time, with no telling what it would do next, and loved because it guaranteed a tremendous spread of Greek culture, still much admired in modern times. They populated its abyssal depth with gods and goddesses; Poseidon, Thetis, Nereus and fabulous monsters. More than two millenia later, these fairy names found their way into an order we call "systematics". The sea-born goddess of love, Aphrodite, became the sea mouse, *Aphrodita aculeata,* Linné. The sea-god Nereus and his beautiful daughters had to come forth to give their names to a whole family of elongated bristle worms; Nereidae, Eulalia, Eunice, Hesione.

The wise elders of Greece freed themselves early from any doctrine, something the Europe of later periods only accomplished after centuries of struggle, and in self-possession one of their philosophers nobly declared: "All life stems from the sea". We know now that life really evolved in the ocean; as did millions of years ago, Stromatolites, primitive algae related to the Cyanophyths, are still building their coral-like structures in tropical and subtropical seas. As Thales of Milet saw it, we still see it today; the sea as an ever rejuvenating fountain of life.

Unfortunately, today this well is threatened by pollution and destruction and, like the tropical rain forest, we will probably never be able to unravel all its secrets.

In the last 20 years marine biology and related sciences has presented us with an immense gain in knowledge about animal and knowledge about marine animal and plant life, and seawater aquarists have a considerable share in this achievement.

What brings on a laymen to devote his persistence, patience, and understanding to one sole group of animals; the giant clams? Most probably the same inquiring mind which drove the Old Greeks and Karl von Linné, or Charles Darwin. The giant clams with their fascinating colourful mantles would be on the brink of extinction without the endurance of the scientists who made it possible to propagate these mussels in captivity. Continuous observation and experimentation were necessary to render larvae into small clams. This way is recapitulated in this book. I hope, together with the author Daniel Knop, that the book will find a wide distribution.

Winterthur, September 1994
Peter Wilkens

Acknowledgments

This book about giant clams is born out of a profound interest for these animals, grown over the years. However, without contributions from many people from home and abroad it would not have been possible to write it. Too many contacts were made in many countries during the writing of the manuscript to mention every one, but they all have my cordial gratitude.

Nevertheless, some of them I must mention by name. The publication of the book would not have been possible without their never tiring help. Others impressed me by the enthusiasm and the international co-operation, prevailing in the worldwide effort to propagate clams. Whenever I have met people of the clam family, I received a warm welcome and firm support. Many friendships developed while working on the manuscript. The experience of an international exchange of ideas between aquarists and scientists has been a great joy. Thus the worldwide preparatory travels and correspondence became unforgettable memories.

I have to thank the Marine Science Institute of the University of the Philippines in Manila for their friendly support, in particular Prof. Dr. Edgardo Gomez and Dr. Suzanne Mingoa-Licuanan for their active help. My good friends Dr. Gil Jacinto and Dr. Sonia Jacinto arranged many important connections. I also have to thank Carmen Belda for letting me use her data on clam nutrition, and the biologists Hilly Ann Roa and Bert Estepa for their detailed information and diving excursions.

I also received support from the ICLARM (International Center for Living Aquatic Resources Management). The ICLARM-Coastal Aquaculture Center in Honiara, Solomon Islands, helped to publish new material about clams in international journals and in addition provided fascinating photo material. Special thanks to Stephanie Pallay, Beckie Nachtrieb, and Mark Gervis.

I also have to acknowledge the Marine Laboratory of the Silliman University in Dumaguete, Philippines, and in particular the director, Dr. Hilconida Calumpong for their support and photographic material. The charming biologists Rio Abdon-Naguit, "Wini" Erwinia Solis-Duran and Grace Ozoa gave me a hearty welcome and helped with instructive discussions and diving excursions.

I have to thank the Marine Biology Office of the University of San Carlos, Cebu, Philippines, for the invitation to sail on their research vessel. This excursion highly contributed to this book and widened my personal horizon. Chona Sister and Jason Young supported my work during the journey. A special thank to them. I also have to thank the director of the Marine Biology Office, Dr. Filipina Sotto for her support and for the fascinating conversations we had.

Dr. John Norton of Oonoonba Veterinary Laboratory in Townsville, Australia, supported me in a particular way by providing excellent photographs and helped investigating the "White Spot Disease". He had great influence on the chapters on anatomy and clam diseases. Furthermore, I am grateful for support from Dr. John Lucas (Australia) and the ACIAR (Australian Centre for International Agricultural Research), for providing me with scientific literature and allowing me to use Dr. J. Norton's beautiful pictures on clam anatomy. ACIAR correspondent Janet Lawrence is gratefully acknowledged for her never tiring efforts to establish contacts.

The artists Melchior Buelo and Henri Rivero of the Biological Faculty of the University of the Philippines provided the excellent illustrations according sketches in literature and my own proposals. Their work is highly appreciated. During our stay we lived with the families of William Garcia in Pangasinan and Mike Lopez in Dumaguete, enjoying their warm hospitality. Grace Lopez showed much interest and invested much time in our work.

At home, generous support was provided by Dr. Hans-Heinrich Janssen letting me use his result on the mussel *Corculum cardissa*. Karl-Heinz Schmitt of "Schmitt Aquarienbau" in Kleinwallstatt diligently built the experimental tanks according to my plans and Stefan Albat took good care of my aquaria during my absence from home. Svein Fosså, Bobby Wong jr., Rolf Hebbinghaus, Klaus Jansen, and Enrico Enzmann provided excellent photographic material. Our friend "Madz", Dr. Madeleine de Rosas Valera established contacts and helped in many other ways. My daughter Melanie helped with the arrangement of the photographs and legends.

Finally I am grateful to my wife Rosalinda for constantly organizing and co-ordinating my travels, contacting people and not avoiding physical hardship in order to support my work. Her's is a substantial effort to create this book.

Daniel Knop

Between Myths and Science – the Giant Clams

The clams of the family *Tridacnidae* with their usually very colourful syphonal mantle are among the most popular animals and are kept with great devotion for many years in seawater aquaria. Their appearance, however, is also very impressive when in their natural habitat, the reef; the fluorescent blue or green coloured mantle of the smallest species *T. crocea*, for example, which sometimes forms large colonies of astounding densities, or the gigantic specimens of *T. gigas,* whose shells can reach a diameter of more than one meter.

T. gigas is the largest of all known clams. The impressive growth rate of the tridacnids is likely to be due to their ability to cultivate plants in their body tissue. These plants comprise of unicellular algae of which the metabolic products add to the filter food of the clam. While smaller clam species are restricted to suspended food, which is filtered by the gills out of the sea water, the tridacnids have adopted the concept of endosymbiosis from the coral polyps and have unicellular algae in their syphonal mantle. However, in doing so, they have improved this strategy significantly: In contrast to the corals they cultivate these algae in a special circulatory system which enables them to keep a substantially higher number of symbionts per square unit. This circulatory system was discovered only very recently (J. H. Norton et al., 1992).

The concept of endosymbiosis has proved to be beneficial to the giant clam, as it enabled it to settle in many tropical coral reefs from shallow water down to great depths. I found these

A pretty group of clams of the species Tridacna crocea and Tridacna squamosa which I found in breeding cages in the Philippine province of Pangasinan (UP MSI).

A larger specimen of T. gigas. To meet one of these animals is on of the highlights for a diver.

The shells of Tridacna gigas often reach impressive sizes. Lisa Lopez shows here a pair of shells, that was found by chance on her family's premises during digging works. The lenght of these shells was 103.5 centimetres.

We found this beautiful shell of Tridacna gigas at the coastline of Cebu, Philippines.

clams in depths of as much as 18. 5 metres. If the extremely rare species *T. tevoroa* is included, maximum depths of about 30 metres can be reached. The reason for these animals being endangered is not due to insufficient adaptation to their natural habitat, but to dramatic overfishing by the suppliers of the food industry, in combination with increasingly better and more effective catching methods (H. Govan et al., 1988).

The rapid change that the image of the clams have undergone in the Western World is surprising. The people of South East Asia and the Pacific Islands have long appreciated these animals as a delicacy used the large shells and at times paid a fortune for the adductor muscle which the Chinese believed aphrodisiac powers. On the other hand, the image of these molluscs in the Western World for a long time remained a lore in adventure stories. Horror movies and gruesome novels reported dangerous beasts ambushing divers, and

gave descriptions of gigantic jaws starring with teeth which would grab arms and legs. This distorted picture, which is also responsible for the popular name "killer clam", may have

A particularly beautiful specimen of Tridacna gigas. Amongst roughly 4000 specimen, I found only two individuals with this kind of a pattern.

11

contributed to the fact that scientific research noticed these animals relatively late. As early as 1825 the molluscs are mentioned in de Blainville's reports, yet scientists have only investigated the giant clams more intensively and systematically in the last three decades. More than 70 % of the 290 scientific publications on giant clams known today have been written after 1970. In the years between 1900 and 1969, there was not more than an average of one single publication per year.

Sir Maurice Yonge, the great British marine biologist was among the few people who worked intensively with this group of animals. He described their endosymbiosis as early as 1936 and as late as 1980, he still published papers on these molluscs. In 1965 J. R. Rosewater redescribed the family *Tridacnidae* which so far had been in a big muddle, and in his substantial publications built he laid a basis for a better understanding of these extraordinary animals.

Anybody who has worked intensively with giant clams and the research on them for the past 15 years can imagine the enthusiasm and pioneering spirit of earlier generations of researchers. Today, however, our knowledge grows faster than ever before, thus making it almost impossible for a researcher to discover something entirely new.

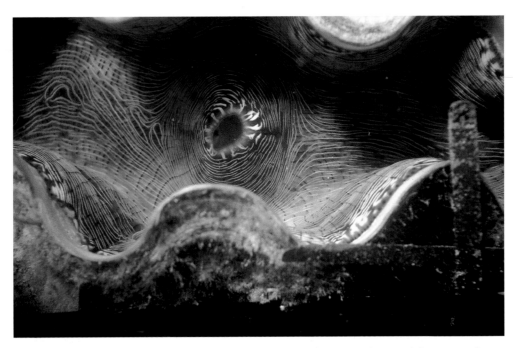

The fluorescent colour patterns of this Tridacna derasa create dummy outlines and thus camouflage the contours of the mantle.

Among the few who share this privilege are the pioneers of clam research. Names like Yonge and Rosewater will always be remembered in connection with these most beautiful and special molluscs, as they form the backbone of this field of research which culminated recently in their successful propagation, and thus saved some of the clam species from extinction. When I contacted biologists who work on the development and optimization of clam propagation, I found it particularly interesting that this field is obviously dominated by women. While many other fields in scientific research are frequented by men, I found mainly women in the field of clam propagation and research. The biologist Dr. Filipina Sotto explains this by emphasizing the role of care in breeding giant clams; brood care motivation would be activated in women which would create a strong emotional relationship. I myself sympathize with this opinion, however, I do not believe that this explanation is sufficient. On three different continents I met the wives of sea water aquarists which felt rather reservedly towards their husband's enthusiasm. Still they would love and care for the colourful clams in the respective tanks. Quite often the wives were the ones who chose the clams in the pet shop. It appears to me that these miraculous animals are particularly attractive to women.

One question often asked is: what is the purpose of the colourful patterns displayed on the mantle of the clams? Unfortunately this question cannot be answered with certainty by anybody up to now. The brownish ground colouration of the mantle is doubtlessly caused by the symbiotic algae, however, the iridescent pigmentation creating all the different patterns in a variety of colours is not directly related to the algae. I suspect that this pigmentation is primarily a means of protection against ultraviolet light, probably against too much light in general. This view is supported by the fact that the pigments reflect part of the incoming light recognized as colour patterns, when we look at these animals. The fraction of reflected light which reaches our eyes does not reach the symbiotic algae.

Although this explains the presence of pigments, as well as their light dependent increase or decrease, it does not explain the various patterns. The patterns are, according to my opinion, a means of disintegrating the outer contour of the mantle. The animals forced to expose their mantle towards the sunlight and thus making themselves conspicuous to predators, developed an unconventional camouflage mechanism. In contrast to many animals who try to match their body colouration with the background, and in so doing making their body outlines invisible, the tridacnids create dummy body contours which are even more brilliant and more marked than the actual outlines and thus fool the predator. The chances of a clam escaping from being eaten and reaching maturity would therefore be dependent on the intensity and variety of its colour pattern. Animals with an easy to identify body outline, would be recognized first and devoured first. However, this hypothesis lacks valid evidence.

Another hypothesis of mine concerns the marked toothing at the upper rim of the shell. These tooth like nodes are plausibly explained in science: they are derived from the vertical folds that stabilize the shell. The nodes are just a progression of these folds. I believe that these tooth structures on the rims of the shells could have evolved into a signal reinforced by the predator's selection. The shells in *T. crocea,* which is the smallest species, are substantially

With their impressively toothed rim these specimen of Tridacna squamosa resemble hungry gorges which might discourage predators.

As soon as the shadow of a passing fish falls onto the group of clams, they retract their mantles and expose their "teeth".

thicker relative to their size, compared to the shells of other species. My hypothesis is supported by the fact that their shells are not folded and still have the nodes on the upper rim. I suspect that the effect is considerable, when the mantle is retracted and the teeth are exposed they can easily intimidate predators. This effect is additionally reinforced by the impression that the shells act like jaws that are closed. In larger specimens of the species *T. gigas* and *T. squamosa,* I often observed that the shells are closed and slowly opened several times in a jerky manner as soon as the mantle is touched, as if they would simulate defensive "snapping".

Additional support for this kind of a signal comes from the fact that similar signals have developed in other clam species. A certain oyster species, for example, seems to use the same defensive strategy. Teeth trigger a strong emotional reaction even in humans, which can be measured in form of a dilatation of the pupils (Eibl-Eibesfeldt, 1971). This reaction also used by advertisements is according to the behavioural biologist Eibl-Eibesfeldt an innate response of the tooth pattern of a predator. I am convinced that similar behavioural signals are common throughout the animal kingdom. In the family *Tridacnidae* these have led to a defensive strategy using a morphology originally derived from a character for strengthening the shells.

But whatever theory we like to believe, the fact remains that the reflecting pigments of these animals create an incredible opalescence almost unique in the coral reef and that the appeal of these peaceful clams has therefore attracted many sea water aquarists from all over the world.

Species Descriptions

Introduction to Species Identification

In this chapter all nine scientifically described species of the family *Tridacnidae* are introduced. The description of each of the species usually provides a means of precise allocation of the life animal or its shells, provided the typical characters of the species are fully expressed in the specimen to be identified. The inexperienced reader is advised to rather concentrate on the shells than on the colour patterns of the syphonal mantle, when trying to identify a clam. Although the colour patterns and the shape of the mantle alone usually allow the expert a precise allocation, a lot of routine is needed in some of the species. Using colour patterns alone can lead to false identifications as the patterns are highly

A group of particularly colourful clams of the species T.crocea and T. maxima in a coral reef tank.

variable. In contrast, the shells usually allow even the inexperienced observer to differentiate between species at first sight.

However, it can be tricky to identify very young animals, as the shells of a number of species closely resemble each other. Although the typical

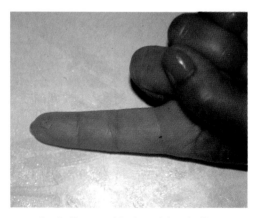

Juvenile shells resemble the adult's shells quite well. This photo shows a small shell of the species T. crocea.

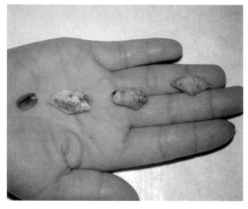

At a length of 18 mm species identification is usually no problem. From left to right: T. crocea, T. gigas, T. maxima and T. derasa.

15

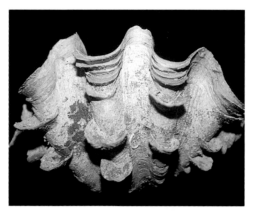

The shell of a T. squamosa with a length of 18 cm. While the distances between the rows of scales at the middle of the shell indicate a normal growth rate, the denser rows at the upper rim indicate that the shell developed much slowlier in the aquarium (caused by a lack of hydrogen carbonates). The time of change of the environment from the sea to the aquarium is clearly visible.

A collection of clam's shells in the Marine Science Institute of the Philippines in Pangasinan. This collection comprises seven of the nine known Tridacna species. (T. crocea cannot be seen in the picture.)

characters of the respective species may be present, they can be expressed too weakly to allow identification, even by an expert. For instance, the pairs of shells of juvenile T. crocea are impossible to tell from those of the same sized T. maxima. Similar problems arise when comparing the shells of T. gigas and T. derasa if their length is less than one centimetre. However, animals of such small sizes are not available in the aquarium trade and they will probably never be, as they have a high mortality rate when kept in closed systems.

In larger clams these kinds of problems of species identification mainly emerge in T. maxima, because this species often develops shorter and higher shells, which are also very symmetrical and resemble those of T. squamosa. Such an example is given in the photographic material.

The other species are usually easy to identify. However, if a clam is raised in a tank, the artificial light conditions and other artificial factors can influence the growth and thus the shape of the shell. Aquaria raised T. squamosa for example, occasionally show much denser rows of scales. Other aberrations in shape caused by an adaptation to aquarium conditions are also possible and should be taken into account when identifying a species.

In order to provide a direct possibility for comparing the most prominent characters for identification and to make identification easy for the inexperienced reader, I have summarized the most obvious traits of each species in a table at the end of this chapter. In contrast to scientific tables, which require the differentiation and precise measurements of the shells of dead animals, my table also includes the mantle of the clam, and therefore it can be used for live clams in the pet shop, for example. To make the use of the table as easy as possible, I concentrated on the most typical characters of each species. As long as the characters are prominently expressed, this facilitates species identification, but it makes it more difficult to identify a species if shells are atypical or not fully expressed. In critical cases this table might therefore prove to be insufficient.

The rare species *H. porcellanus* and *T. tevoroa* are also included in the table, which does not mean that they are available in the aquarium trade. It seems to me, however, that breeding efforts undertaken in the hatcheries may in the long run make these species obtainable. The table is meant to enable divers and sea water aquarists to identify all living clams in the reef, as well as provide a tool to recognize wrongly identified species in the aquarium literature.

The Outer Characters of the Giant Clams:

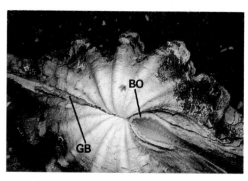

View from below:
GB = outer ligament (inner ligament not visible),
BO = Byssus opening

▶

Outer characters of giant clams:
ML = mantle (syphonal mantle),
E = inlet, A = outlet, Z = teeth,
S = scales (not present in all species),
VF = vertical folds

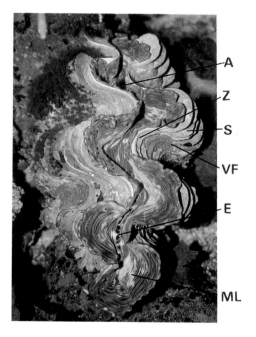

The complete biological classification of the giant clams:

Tribute: *Mollusca*
Class: *Bivalvia (mussels)*
Order: *Veneroidea (venus mussels)*
Family: *Cardiacea (heart shaped mussels)*
Subfamily: *Tridacnidae (giant clams)*
Genus: *Tridacna and Hippopus*
Species: *T. gigas, T. derasa, T. maxima, T. crocea, T. squamosa, T. tevoroa, T. rosewateri, H. hippopus, H. porcellanus.*

The shell of Tridacna crocea is relatively smooth and hardly has any vertical folds. Still the teeth are marked at the upper rim.

Species

Tridacna crocea (Lamarck, 1819)

Tridacna crocea is the smallest of all Tridacna species. Even fully grown, its shells do not grow larger than 13 to 15 centimetres. They are usually smooth and finely corrugated. At the zone on the upper rim of the shell where the shell grows, a number of rows of thin scales are visible, which are very fragile

Tridacna crocea has the largest byssus opening compared to its shell length.

Typical appearance of a Tridacna crocea. The lower middle segment of the mantle is particularly narrow in this species.

Tridacna crocea can bore its shells almost entirely into the limestone. In the centre of the rock in the picture above you can recognize the hole which the clam left behind.

Various growth stages of Tridacna crocea.

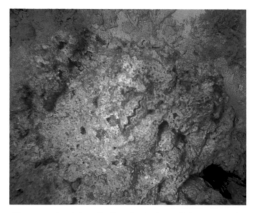

Close to the island of Bohol I found this pair of shells of a deceased Tridacna crocea in three metres depth, which stuck in the ground by two thirds.

and easily break off. The upper rim of the shell occasionally shows a yellowish or orange-yellow band, which is exposed when the syphonal mantle is retracted. In contrast to other species the shells of this species can be closed almost completely. In *T. crocea* the byssal orifice on the bottom side of the clam is very large.

The species lives in colonies in the upper light exposed litoral. Due to their small sizes, these colonies sometimes reach surprisingly high population

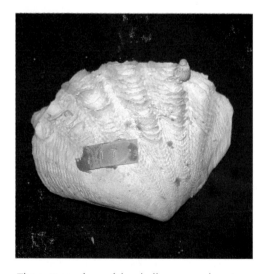

The outer surface of the shell appears almost like a file due to its fine horizontal corrugation. The shells are short and very high.

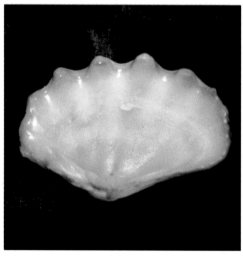

An orange band around the inner rim is not rare in Tridacna crocea.

Tridacna crocea in an aquarium

A colony of young Tridacna crocea. The synchronized closing of the shells when a shadow falls on them, is a fascinating observation as the clams display a jerky retraction of their mantle.

densities. At the Great Barrier Reef of Australia, for example, more than 200 individuals have been counted living in a single square metre (Hammer, 1978). What I found particularly interesting, was that T. crocea literally bores itself into the ground. This behaviour has given the clam the popular name "boring clam". For boring, the animal uses its fine strong corrugations, which give the shells an almost file like appearance. In fact, the chalky substrate is filed off by jerky contractions.

This is achieved not only by mechanical action, but also by a chemical trick: In the vicinity of the byssal orifice, the animal produces organic acids which have a dissolving effect on the chalk. This softens the limestone and thus enables the shells to penetrate. By the time the animal sinks into the protective substrate and only the rims of the shells, or, when opened, the mantle is visible. This mode of settling does of course require very

strong shells. This is why the shells of the smallest Tridacna species are relatively thick and heavy.

The syphonal mantle of T. crocea is dominated by brownish ground colours. Blueish iridescent spots or various patterns of other colours are scattered over the mantle. Some of the most beautiful animals are really very colourful and display green, blue, beige, purple or orange spots. In a few rare cases the mantle is covered fully or partly in a metallic turquoise colouration, which makes those specimens especially attractive gems for the hobbyist. Unfortunately these colour patterns depend highly on the light conditions and tend to fade when those are altered. After some time in a tank the animals may lose the intensity of their colouration. This is due to the fact that the reflecting colour pigments are part of a light protection mechanism which adapts continuously to the incoming light intensity and spectrum.

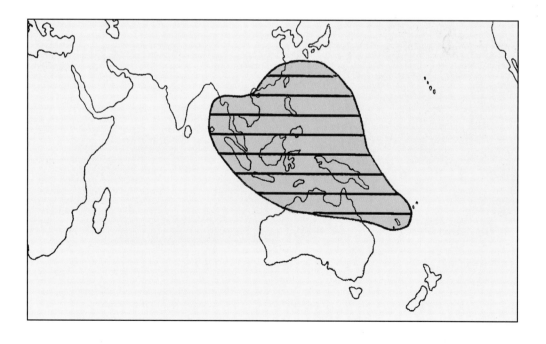

Distribution: *T. crocea* is widely distributed in the Indo-Malayan and Western Pacific ocean, from Thailand to New-Caledonia.

Tridacna squamosa (Lamarck, 1819)

Tridacna squamosa is a much larger species reaching shell lengths of 30, occasionally even 40 centimetres. The most important features of *T. squamosa* are the obvious rows of scales, from which the name "squamosa" derives (squama = scale). These rows are not only present on the upper rim of the shell, but it stretches from the bottom to the top over the entire shell. The denser the shells at the upper rim, the longer and wider they are. Normally these scales with their pockets, provide shelter for a variety marine organisms. As in other clam species, sponges and other invertebrates, or even small crabs can be found in them.

The colouration of the shells is variable and can be white with a lemon yellow band at the upper rim, or entirely yellowish with orange spots. Occasionally the whole shell is yellow or orange to pink. Shells of these colours have achieved top prices in the Philippine shell trade. The upper rim can be closed densely in this species. Seen from the side, the shell resembles an almost perfect symmetrical triangle.

The syphonal mantle of *T. squamosa* is usually speckled in colour giving the

A formidable specimen of Tridacna squamosa in the aquarium of Klaus Jansen. The colouration of the mantle is typical. You can recognize, that the middle part of it, which contains the two syphonal openings, is significantly wider than the one of T. crocea. The ring of tentacles, which borders the inlet in T. squamosa is clearly visible. At the bottom left you can see a typical specimen of T. maxima.
Photo by K. Jansen

Tridacna squamosa has a very typical shape, which is easy to tell from the other species.

The scales stretch from the upper rim of the shell to the bottom. The distances between the rows of scales are relatively wide.

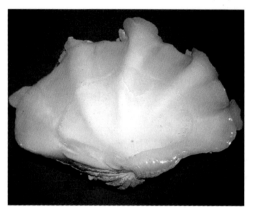

In T. squamosa the almost perfect symmetry is very obvious. It serves as an important character to distinguish the species from T. maxima.

The same symmetry can be seen on the inside of the shell.

Tridacna squamosa in an aquarium. The picture shows an almost adult clam.

animal a colourful appearance with green, blue, brown, yellow and orange spots and bands.

T. squamosa lives on the reef down to depths of 18 metres, on living or dead stony corals, commonly in community

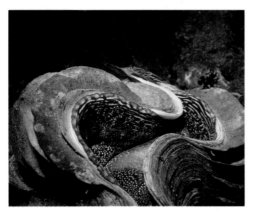

The picture shows a Tridacna squamosa, grown in an aquarium with a shell length of 32 cm. In comparison to the shells in the former pictures, the clam shows a much narrower distance between the rows of scales. This is probably due to less favourable growing conditions. At the upper rim of the shell you can recognize the growing zone with its newly formed calcareus substance.

The impression of teeth is enhanced by the widely spreaded scales in a closed Tridacna squamosa.

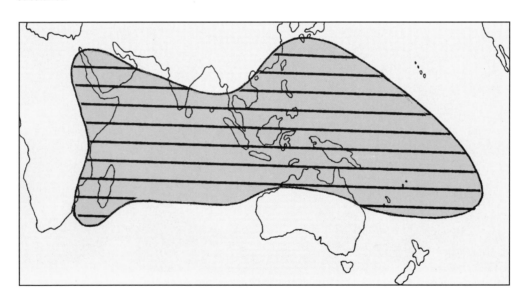

with stony corals of the genus Acropora. To settle, the clam attaches to the substrate with its weak byssal filaments.

Distribution: from East Africa and the Red Sea to Polynesia.

Tridacna maxima (Röding, 1798)

Tridacna maxima resembles *T. squamosa* in size, reaching a maximum shell size of 30 to 40 centimetres, but in general it remains considerably smaller. In contrast to *T. squamosa,* the shells of *T. maxima* are not symmetrical, but distally prolonged. This has given the species the synonym Tridacna elongata.

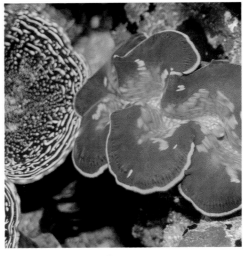

A relatively inconspicuous specimen of Tridacna maxima in an aquarium.

T. maxima also has rows of scales, but these are confined to the vicinity of the upper rim. The lower parts are smooth and, if at all, slightly corrugated. The rows of scales are considerably denser in *T. maxima* than in *T. squamosa.* Shell colouration, on the other hand is very similar to *T. squamosa* varying from pure white to yellowish or orange.

A beautifully coloured specimen of Tridacna maxima.

Like *T. crocea*, *T. maxima* bores into the ground, however, not as deep as the former. It lives in large colonies. In Polynesia, for example, more than 60 individuals were seen per square metre in some of the coastal stretches (Richard, 1985). In such dense populations, the underground is hardly visible, when all the clams are open, as they use each and every square centimetre of the substrate for their mantle. They settle slightly imbedded

Shell of a typically shaped Tridacna maxima with a length of 10 cm. The shell is asymmetrically shaped and the ligament is extremely short.

A view onto the pair of shells from above reveals the high number of vertical folds which is occasionally found in Tridacna maxima. In many specimen I counted up to seven folds and teeth.

This picture shows two differently shaped shells of Tridacna maxima. At the upper rim of the upper shell the characteristically dense rows of scales are clearly visible. The lower shell shows a more flat elongated shape.

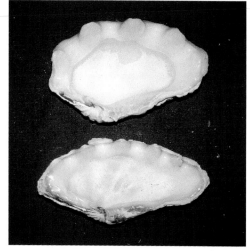

A view at the inner side reveals the obvious asymmetry of the shell. If you imagine a middle axis running through the deepest point of the shell, you will see that the shell is unilaterally elongated. This gave it its synonym Tridacna elongata.

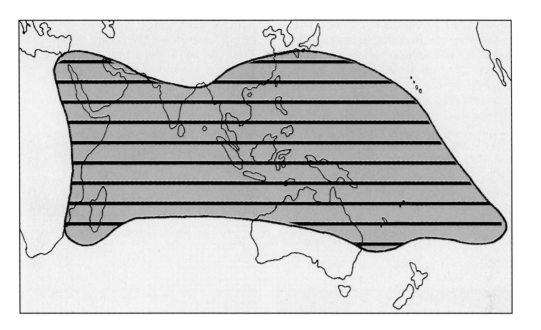

into the substrate or attached to dead corals with their byssal filaments.

Distribution: *T. maxima* has the widest distribution of all clams stretching from East Africa and the Red Sea to Polynesia

Tridacna derasa (Röding, 1819)

Tridacna derasa is the second largest species among the clams. The shells reach lengths of 50 or even 60 centimetres in rare cases. Instead of a byssal orifice the clam has only a narrow gap. The teeth at the upper rim are rounded and the shells close perfectly, which is a clear difference from *Tridacna gigas*.

The two species are often confused, particularly when in their juvenile stages while their species specific characters are not yet fully developed, and the size

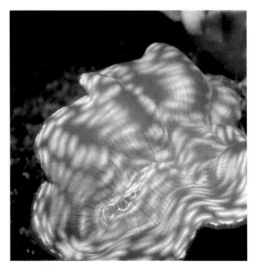

The very typically coloured mantle of an aquarium raised Tridacna derasa. The rim usually has a blue fringe while the area of the mantle is dominated by fluorescent golden yellow, as well as blue and green colours.

27

of the animal gives no hint either. The colouration of the mantle may help in these cases. *T. derasa* has almost always a very colourful mantle dominated by bright iridescent blues and greens, while *T. gigas* is usually brownish in colouration. In *T. derasa* you can normally count six or seven vertical folds in the shell, whereas *T. gigas* has only four or five folds.

Another species often mixed up with *T. derasa* is the rare species *Hippopus porcellanus* which has a smooth shell that resembles the one of *T. derasa*.

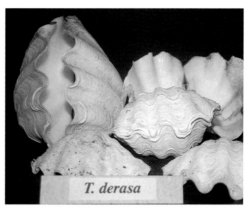

The shells of Tridacna derasa are shaped in a very special manner and therefore easy to identify.

The shell of Tridacna derasa usually possesses six vertical folds. This is the criterion to separate it from Tridacna gigas which looks very similar when young.

The almost perfect symmetry is also obvious when looking at the inside of the shell.

Even when still only a few millimeters long, the shells have a very characteristic shape.

A group of juvenile Tridacna derasa with a shell length of about 35 cm.

28

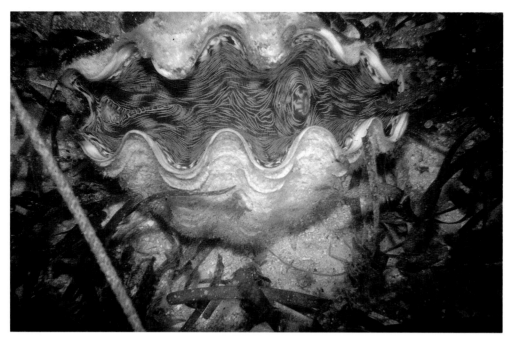

A miraculously patterned Tridacna derasa in the sea with a shell length of about 40 cm. Besides the unusually beautiful colouration the strongly curved shells are conspicuous in this specimen.

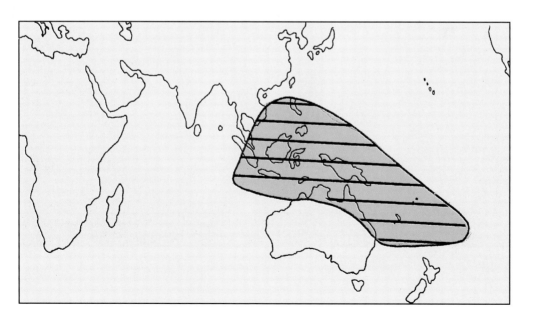

29

However, the two can be distinguished by specific characters in the region of the byssal orifice, which will be mentioned in the description of *H. porcellanus.*

T. derasa lives mainly on the outer side of the reef in depths between 4 and 20 metres.

Distribution: Australia, Philippines and Indonesia.

Tridacna gigas (Linné, 1758)

Tridacna gigas is the largest species of clams. This impressive animal which can reach a size of more than one metre, quite regularly was given the well-known synonyms "giant clam" or "killer clam". Even very large animals are not rare in places where the clam is still found.

The inner side of a Tridacna gigas (same as below). The length of this shell was 94 cm. However, when I looked at it more carefully I realized that it has lost some of its original size which must have been more than one metre.

Shell of a Tridacna gigas which is displayed in front of a local museum in a Philippine township.

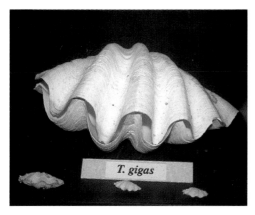

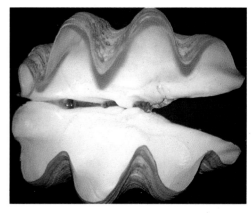

Adult Tridacna gigas are the only giant clams unable to close the upper rim of their shells completely. Even when closed, a part of the mantle is always visible.

The very long and conical curves of the upper rim are typical character of this species.

However, the real sizes of the clams are often exaggerated and stories are told about animals reaching sizes of 150 centimetres and more. This seems a bit too much as the largest known specimen was found to measure 137 centimetres and was discovered around 1817 on the north western coast of Sumatra (Tapanuela). The weight of the two shells was 230 kilograms which suggests that the live weight of this animal must have been roughly 250 kilograms. Today these shells are under the curation of a museum in Northern Ireland (Arno's Vale). Another huge specimen was found in 1956 off the Japanese island of Ishigaki, but was not examined scientifically before 1984. The size of its shells was 115 centimetres and the weight of the shells and the soft parts

Tridacna gigas has only four or five vertical folds. This is the main criterion to separate it from the very similar shell of T. derasa. Long scales are not existent in these shells.

Due to its symmetrical and well balanced shape the shell of Tridacna gigas can easily be differentiated from shells of T. maxima and others.

was 333 kilograms. The scientists estimated its live weight to be 340 kilograms.

A larger specimen of Tridacna gigas in the middle of a sea weed bed. I would estimate the length of its shells to be about 70 cm.

T. gigas has a certain similarity with *T. derasa* which is particularly true for younger specimens when the typical characters have not yet fully developed. One of the criteria for differentiation is the inability of *T. gigas* to fully close its shells at the upper rim. However, this can only be recognized with increasing age and growth. The teeth do not close as perfectly as in *T. derasa*. Small gaps always remain between the shells through which the retracted brownish-yellow mantle can be seen. Another criterion is the number of vertical folds found on the shell. *T. gigas* has four or five folds, whereas *T. derasa* has six or sometimes seven folds. A byssal orifice is only allusively present in *T. gigas*.

T. gigas lives in the flat coral sand or broken coral, but can also be found at a depth of as much as 20 metres.

Distribution: *T. gigas* is not one of the rarest, but still it is one of the most endangered species amongst the clams. It has become extinct in large parts of its former range which stretched over the entire Indo-Pacific. Even in places in which it still exists, populations are diminishing quickly. The main reason for this is likely to be the intensive exploitation by the mussel catching vessels which provide the Asian gastronomy with the meat from the

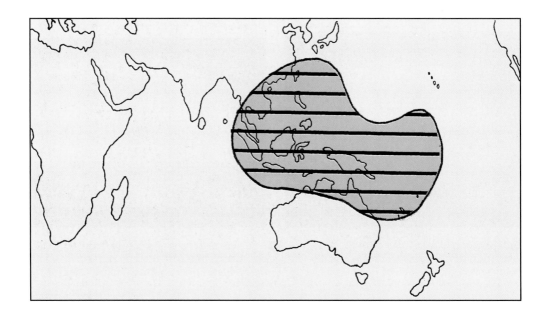

muscles of this animal. Mainly the large, adult animals are killed as they are most profitable. Unfortunately these adults are vital for the recruitment of offspring. Thus, not only the populations of this strange and beautiful reef dwellers are reduced, but also their recruitment is disturbed, so that exploitation cannot be compensated.

Tridacna tevoroa Lucas, Ledua and Braley, 1990 (T. mbalavuana, Ladd)

Besides *Tridacna rosewateri*, *Tridacna tevoroa* is the rarest of all clams. It was

▶

A specimen of Tridacna tevoroa. The minor quality of the photographic material of this giant clam is due to the fact that these animals are extremely rare and that photos are often taken under difficult circumstances.

Photo: Marine Lab

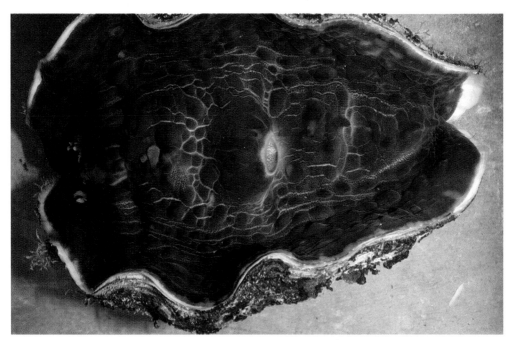

Tridacna tevoroa, the so-called "devil's clam" is very different from other clams due to its dark warty syphonal mantle. Photo: Dr. John Norton, Australia

33

A lateral view of the shell of Tridacna tevoroa suggests a similarity with Tridacna derasa.

When looking at the inner side of the shell it becomes obvious that the pallial line (see chapt. "anatomy") at which the retracting muscles of the mantle are connected with the shell is very distant from the upper rim.

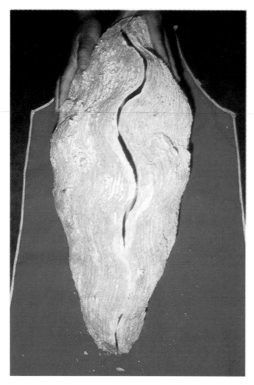

The typical appearance of Tridacna tevoroa is due to its flat shells and the soft curves of its teeth.

discovered only very recently. Its unusually flat shells reach sizes of 50 centimetres or more. The overall appearance of this animal is so special that it can easily be distinguished from other species. The shells which are only very slightly curved, develop hardly any

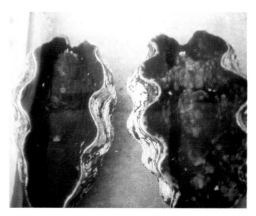

The curves of the shell can vary quite a bit within the species. Occasionally you can find specimens with shells which are more arched.
Photo: Silliman University, Dumaguete, Philippines

34

teeth. Only very flat folds can be recognized, which fit exactly into each other and thus the animal can close its shells completely. From the side, the shell resembles the smooth shell of *T. derasa*, but cannot be mixed up with it as it is hardly curved. Close to the ligament one can find a dark red band

pattern which facilitates identification even if there is only one shell available.

Another criterion which characterizes *T. tevoroa* and two well-known *Hippopus* species is the lack of an overhanging mantle. While in all other Tridacna species the syphonal lobe is

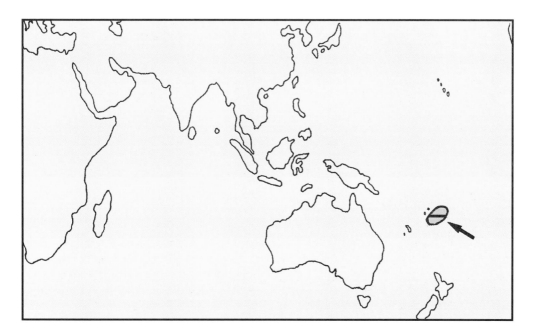

protruding widely over the rim of the shell, the mantle of *T. tevoroa* (and the two *Hippopus* species) ends exactly at the upper rim of the shell.

T. tevoroa lives in much greater depths than the other clam species. They are found at the outer reef zones in depths between 14 and 30 metres.

Distribution: restricted to the island of Tonga and the Eastern Fiji-Islands.

Tridacna rosewateri
Sirenko and Scarlato, 1991

Tridacna rosewateri was described in 1991 by B.I. Sirenko and O.A. Scarlato. The holotype of the species (the

The vertical folds are developed extremely in Tridacna rosewateri. Probably this is a measure to stabilize the very thin and light shell.

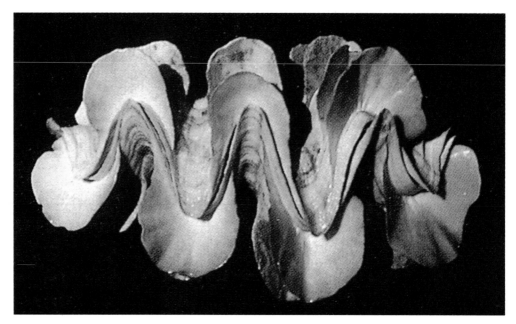

Holotype of Tridacna rosewateri. This photo from above shows clearly the specialties of this clam: The scales which protrude almost at right angles to the shell are longer and wider than those of Tridacna squamosa. The length of the teeth on the upper rim of the shell resemble those of Tridacna gigas. All four photos of this species were kindly provided by "La Conchiglia", Rome.

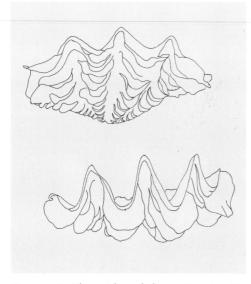

T. rosewateri from side and above. Drawing: M. Buelo after scetches from "La Conchiglia".

specimen used in the first description which has to be accessible to the public in a museum or another type of collection) and eight other specimens called paratypes, are found in the Zoological Institute of Leningrad. Another paratype is stored at the State University of Moskow.

T. rosewateri is an example of adaptation to very special environmental demands. For instance, if heavy shells are no more beneficial in a certain environment, probably because there is no strong current, the animals develop thinner and lighter shells, as the selection pressure for heavy shells is not present. The accompanying lack of stability must be compensated by other measures. In the case of T. rosewateri this is achieved by a strengthening of the radial folds with vertical corrugations that give the clam shell its stability. In

The same shell from below. The marked byssus aperture suggests that the clam is tightly attached to the ground.

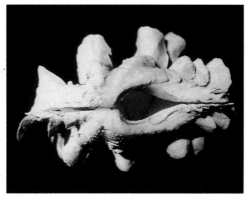

A paratype of Tridacna rosewateri which demonstrates the same typical characteristics.

fact the radial folds are particularly strong in *T. rosewateri*.

In addition, the very long scales of this species might be part of a defensive measure to intimidate predators. The scales are even more strongly developed in *T. rosewateri* than in *T. squamosa.* Another measure of defense, might also

be the very long teeth that are formed by the upper rim of the shell.

The size of this species is rather modest: the lengths of the ten shells so far measured range between 67 and 191 millimeters. The average length is 150 millimeters and thus are about the size of an adult *T. crocea.* The well

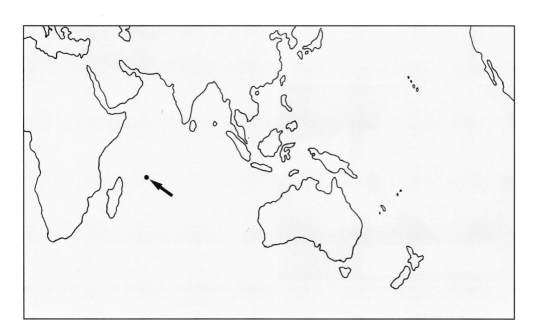

developed byssal orifice at the bottom of the shell suggests that a bundle of filaments holds the clam tight to the ground. The editors of the scientific journal "La Conchiglia" were friendly enough to provide me with the photos of the shells, which stem from the only publication on this species up to now.

As has been indicated above, *T. rosewateri* developed in a geographically very isolated environment. Due to the shape of their shells, I suspect that the species might have derived from a common ancestor with *T. squamosa*. Its status as a true species is not yet established, as it was not possible so far to separate it from *T. squamosa* and *T. maxima* with which it has a number of characters in common. Maybe one day we will be able to examine the soft parts of this animal of which we have nothing but the shells so far. Only then we will have the material for an exact biological classification.

Hippopus hippopus (Linné, 1758)

Hippopus hippopus is significantly different from most other species of the family. The very special shape of its very thick and heavy shells, which are strongly arched and have a rhombic outline from a side view, has given the animal a number of popular names in the Far East. It is called for example, "horse hoof elam" or "bear paw clam". The scientific name also

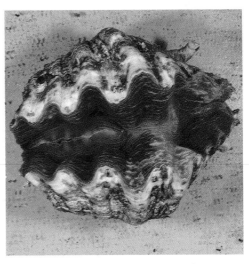

Hippopus hippopus has a very special appearance, which is different from Tridacna species in a number of traits.

The view from above onto the closed shells shows their strong arch.

derives from these synonyms (hippos = horse, pes = foot). Very characteristic are the red horizontal bands on the outer side of the shell. These coloured bands give the shells an almost red appearance when approached from a distance, and gave the animal the popular name "strawberry clam". The maximum size of the shells is about 40 centimetres.

The largest Hippopus hippopus shell ever found in the world. It is housed in the Marine Science Institute of the University of the Philippines (UP MSI) in Pangasinan.

Another speciality already mentioned above is that the mantles of *Hippopus* species like those of *T. tevoroa* do not overhang the shells. When open, the outer rim of the mantle ends with the outer rim of the shell. Thus, in this species the three parts of the mantle consisting of an inner part with the two syphons and the two outer parts, which is typical for almost all Tridacna species, does not exist and the mantle comprises an uninterrupted piece of tissue. The

The red stripes on the outside are a typical characteristic of Hippopus hippopus, which does not occur in any other species.

The inner side of the shell also shows a special shape. Due to its rhombic contour line Hippopus hippopus has the popular name "horse shoe clam".

39

The same pair of shells from above. I measured a shell length of 53.5 cm for this specimen. Usually this species does not grow larger than 40 cm.

From below the closed teeth of the byssus can be seen. This kind of toothed byssus closure is found exclusively in the two Hippopus species.

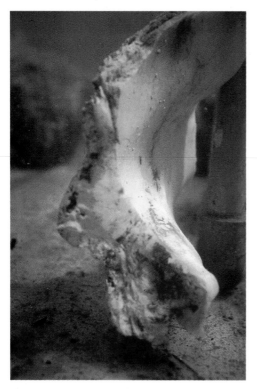

A dissection through the shell of a Hippopus hippopus shows the strong parts on the bottom of the shell which gives the clam its centre of gravity. This helps the animal to stand upright again after it was turned over by strong currents.

Hippopus hippopus clams in the sea. The species is fascinating, not because of its colourful mantle, but because of its special shape. In addition it is according to my experience one of the most robust representatives of the family Tridacnidae. I was lucky enough to be able to import 20 specimens of this species to Germany in 1994, for the first time.

incurrent syphon is not fringed by a ring of tentacles as is common in other Tridacna species.

The byssal orifice is narrow and resembles a slot. Here a feature can be found which is characteristic for the representatives of the genus *Hippopus:* The outer rim of the byssal orifice has clearly visible teeth, whereas all *Tridacna* species have smooth rims. (though the innermost part near the umbo may have small ridges).

The syphonal mantle of *H. hippopus* is usually very inconspicuous. It has a greenish or brownish ground colouration with lighter line patterns, but can also occasionally be yellowish or grey. It is rather dull compared to the *Tridacna* species.

H. hippopus lives in the reef down to a depth of six metres. It prefers the flat sandy stretches or adjoining sea weed beds.

Distribution: *H. hippopus* is common in the Indo-Pacific off Thailand up to Vanuatu. At Tonga, Fiji and Samoa, the species has become extinct.

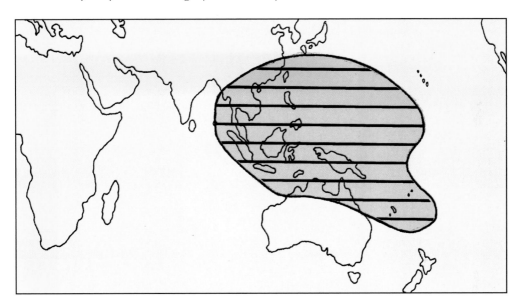

Hippopus porcellanus Rosewater, 1982

Hippopus porcellanus is closely related to *H. hippopus,* and thus there are a number of similarities in the shape of the two species. *H. porcellanus* also has no overhanging syphonal mantle. When open, the outer rim of the syphonal lobe ends with the outer rim of the shell. Due to this very obvious character and the quite typical shape of the shells, identification of the species is normally easy. The maximum size of the shell is 40 centimetres.

In contrast to *H. hippopus, H. porcellanus* has a much smoother shell without any scales. This is the main criterion to differentiate it from *H. hippopus,* but it makes it easy to mix it up with *T. derasa.* The strength of the shell and thus its weight, as well as the weight of the whole animal, is much less than in H. hippopus. Altogether the mussel resembles porcelain very closely. This is reflected in its scientific name as well as in its popular names: in its natural habitat, people call it the "china-clam" or the "porcelain-clam".

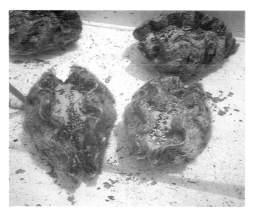

The species Hippopus porcellanus is clearly strongly related to Hippopus hippopus.

The rounded teeth on the upper rim and the flat vertical folds of this species resemble those of T. derasa with which it is mixed up quite often.

Like its close relative *H. hippopus* the much rarer *H. porcellanus* has a very narrow byssal orifice, which possesses teeth perfectly corresponding each other. This "byssal slot" therefore resembles the corrugated upper rim of a closed shell. The mantle of *H. porcellanus* is more conspicuous and colourful than the one of *H. hippopus*, still an olive green ground colouration is dominant. Differentiation between *H. porcellanus* and *H. hippopus* is usually easy if you compare the incurrent syphon. In *H. porcellanus* the syphon has a ring of papillae and tentacles.

The species prefers the sandy patches of the reef.

The contour of Hippopus porcellanus is slightly rhombic.

Inside view of Hippopus porcellanus. The length-height-relation is noteworthy in this species. The height is almost identical with its length, a phenomenon regularly found in Hippopus species.

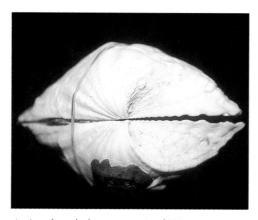

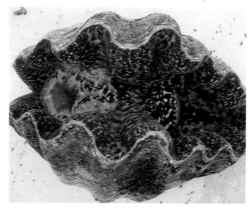

A view from below on a pair of Hippopus porcellanus shells demonstrates the special teeth at the byssus aperture.

This picture and the one below show two colour varieties of Hippopus porcellanus. The ring of tentacles around the intake syphon of the mantle is clearly visible. This is the criterion to separate this species from Hippopus hippopus.

Distribution: *H. porcellanus* is a rare species inhabiting the South China Sea, the Philippines and the Island of Palau. It is the rarest species amongst the *Tridacnidae* with the smallest range, besides *T. rosewateri* and *T. tevoroa*. It is exploited because of its shells which are a prominent article in today souvenir industry on the Philippines (Gomez & Alcala, 1988).

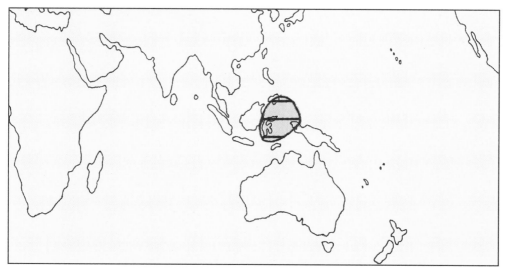

Species Identification Table

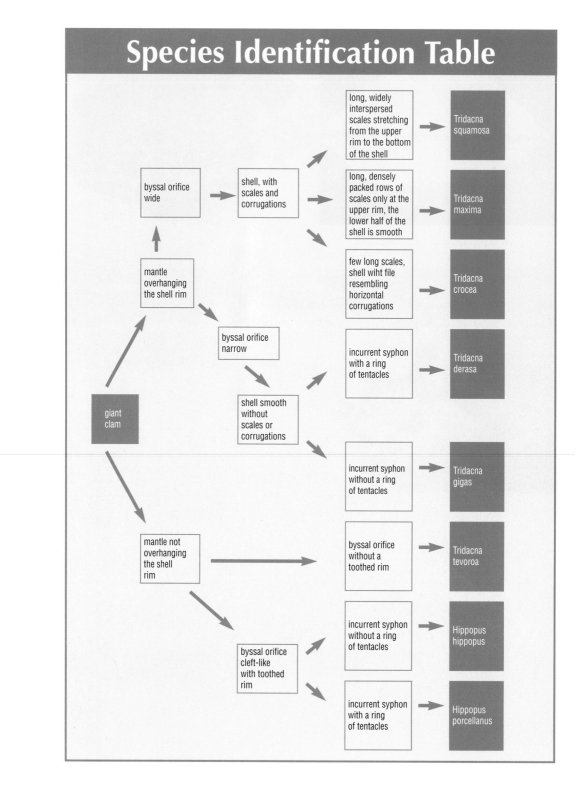

Reproduction

Like all higher developed animals, the clams reproduce sexually. This method of reproduction guarantees that the gene pool between two partners is mixed and thus always new combinations of genes are transferred onto further generations, this way adaptations can take place according to the respective environment.

Because of their special way of life, the clams have to overcome a number of difficulties when reproducing. A higher developed organism, with a sessile way of life - that is, without a chance to move - is quite uncommon. Most of the other sessile animals, like,

for example, the corals have much simpler anatomical structures than the clams. Most of the molluscs on the other hand can move freely, find a mate and deposit a clutch of eggs at a suitable place. Obviously this is not an option for a sessile mussel which has to find alternative strategies to ensure its reproduction. The solution to this problem which has evolved in these animals is called broadcast spawning, the release of eggs and sperm into the open water. This has a big disadvantage, because most of the eggs and sperm are eaten by the numerous hungry mouth of the coral reef. The clams compensate for these losses by producing extremely

Sperm release of a Tridacna squamosa (shell length about 30 cm) in an aquarium.

large quantities of reproductive products. Clams are like most molluscs "simultaneous hermaphrodites" which enables them to produce a very high number of offspring. Each individual is male and female in one, produces eggs and sperm, and therefore does not have to search for a mate of the opposite gender.

Self fertilization is excluded by nature; simultaneous egg and sperm release is not possible, but only within certain time gaps respectively. Therefore everybody depends on a mate which can be any conspecific. The clams which live mostly in colonies, had to find a way of synchronizing the release of eggs and sperm to ensure fertilization.

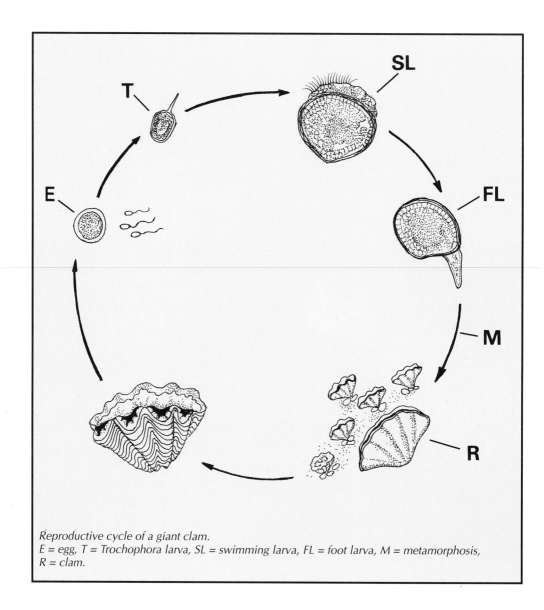

Reproductive cycle of a giant clam.
E = egg, T = Trochophora larva, SL = swimming larva, FL = foot larva, M = metamorphosis,
R = clam.

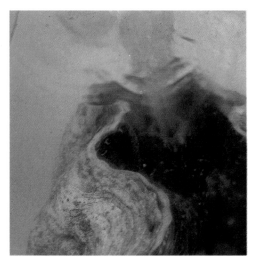

Hippopus porcellanus releasing eggs.
Photo: Silliman University, Marine Laboratory

Tridacna crocea releasing sperm.
Photo: E. Enzmann

The clams solve this problem with the help of a transmitter substance. Although the nature of this substance is not yet fully understood, it is clear that it has hormonal characteristics. The substance is released into the surrounding waters during reproduction by the syphonal outlet. Its scientific name is "SIS" which stands for Spawning Inducing Substance.

Conspecifics discover this substance immediately. The incoming water passes chemoreceptors situated close to the incurrent syphon which transmit the news directly to the cerebral ganglia, a simple form of a brain. Even before the SIS-releasing mussel has started discharging eggs or sperm, initial reactions can be observed in the surrounding animals. The mantle swells in its central region, and the adductor muscle contracts rhythmically.

Soon the clams in the area mean business. One clam after the other fills its water chambers by opening its shells to a full extent. Then the animal closes

the incurrent syphon and contracts the shells vigorously with the adductor muscle, so that the contents of the excurrent chamber is pressed through the excurrent syphon. After a few contractions containing only water, the nervous system of the animal has activated the muscles around the gonads, so that these contract as well and release eggs or sperm into the excurrent chamber. The reproductive cells are then pressed through the excurrent syphon together with the water.

Both egg and sperm excretions comprise the transmitter substance SIS, which ensures that the other animals are not terminating their germ cell release too early. However, as this substance has not yet been isolated, it remains unclear whether the substance is part of the excretions or whether it is attached to the surface of the germ cells.

In the beginning of sperm secretion, the excretion has a flake like appearance, as the sperm cells are stored in dense packages in the gonads. After a number of rhythmic ejaculations following each other in gaps of 15 to 25 seconds, however, the excretion becomes more and more liquified and appears as a homogenous sperm cloud.

The increasing concentration of the transmitter substance SIS in the water usually causes the other animals to join in the "joyous happening". For example, if you would observe a colony of some 30 mussels, it could well be that two thirds of the members are participating in this kind of action. Every three or four seconds, somewhere in the reef, a little cloud emerges, ejaculated by the clams in almost orgasmic convulsions.

Surprisingly small mussels whose shells are hardly half the maximum size are also participating in reproduction. They reach their sexual maturity very early in life, depending on the species, within a few years. Yet, when small, they can only release sperm and not eggs. The ability to also release eggs is reached some time later when the individual is much bigger. Then the clam has become a simultaneous

hermaphrodite combining the abilities of both genders: sperms or eggs can be released within certain time constraints.

If the mussel colony is observed for a while on the reef during reproduction, one can see a jet of excretion from one of the syphons which is quite different from the others. It does not contain the liquid homogenous sperm secretion, but a cord of very fine milk white spherical eggs. These female germ cells have a diameter of 100 micrometer (1/10 millimeter) and are distributed very quickly over the colony by the current. The substance SIS which is released at the same time by the mussel activates the gonads of the other animals. Usually an egg-release initiates the whole process of reproduction: a fully grown sexually mature animal releases its eggs and causes frantic reaction in the colony. After some time more and more animals are participating and a "miniature orgy" develops. 15 to 20 minutes later, the action dies down in some of the animals, but as soon as a new clutch of eggs appears, everybody vigorously starts again.

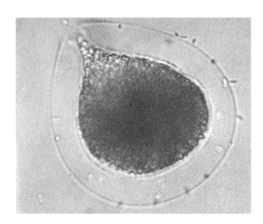

Mature egg before fertilization.
  Photo: Silliman University, Marine Laboratory

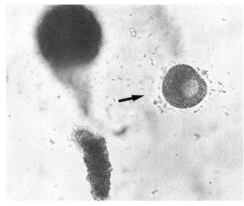

Fertilization of the eggs. The sperm cells can be seen around the egg.
  Photo: Silliman University, Marine Laboratory

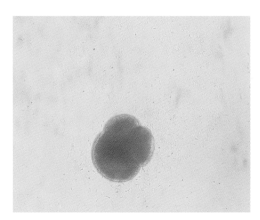

Fertilized egg after two cell divisions.
Photo: Silliman University, Marine Laboratory

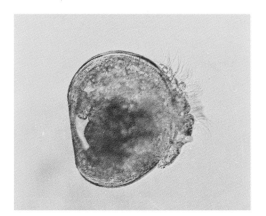

Swimming larva at the age of 5 days. The calcareus shells are clearly visible.
Photo: Silliman University, Marine Laboratory

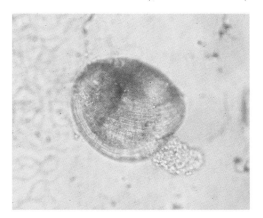

Larva during metamorphosis.
Photo: Silliman University, Marine Laboratory

Later the reef calms down, and only here and there can one see a sperm cloud emerge, while the egg cells, which had made the water turbid, have long been washed away. They carry the genes of the parents in all directions of the wide ocean, yet only very few will ever arrive and settle on a reef. By far the most will be eaten by all sorts of predators during their journey.

The enormous losses are compensated for by enormous production: A single adult clam of *Tridacna gigas* can release more than 500 million eggs during a single mussel orgy. Two animals thus produce occasionally more than a billion larvae. If you consider the high life expectancy of the clams and the fact that adult animals hardly have any enemies, it becomes very clear that even a minimal percentage of successful larvae can ensure the survival of the species.

On their way, however, the larvae have to master a number of dangerous situations. Firstly the fertilized egg undergoes a planktonic stage where it is washed through the ocean by the current. During this time it runs through the stages of a typical mollusc embryonic development. The egg floats in the sea for about twelve hours, until eventually a larva hatches which is called a trochophore larva. The larva immediately starts to produce minute chalk shells. Two days after fertilization it measures 160 micrometer. At this stage it does not have symbiotic algae and therefore depends completely on plankton.

Soon the larva develops a foot, and enters a new developmental period. The foot is used to move on the ground, yet from time to time the larva also swims around in search of the appropriate habitat. At the age of roughly one week, it finally settles on the ground. Although the animal, which measures by now a fifth of a millimeter (200 micrometer), is attached to the surface by byssal filaments, it still changes its location frequently within the first couple of weeks. The mussel turns out its foot between the two lower shell rims and retracts it to the inside when moving.

In contrast to the corals, which are provided with their symbiotic algae by their mother, the little clam has to find its own symbionts (e.g. Stephenson, 1934). Free floating zooxanthellae, drifting in the water as part of the plankton, are taken in by the larva while filtering food. In contrast to the other food particles, however, these microscopic algae are not metabolized, but pass from the intestines into the tissue of the larva which later develops into the syphonal mantle. The entire process is not yet fully understood (Heslinga & Fitt, 1987).

Our little larva has to undergo substantial changes in body architecture, because in the beginning it lives like any ordinary mussel without symbiotic algae. The process of change, described by Yonge as early as 1936 which is scientifically named ontogenetic rotation process leads to a complete rearrangement of the bodily design of

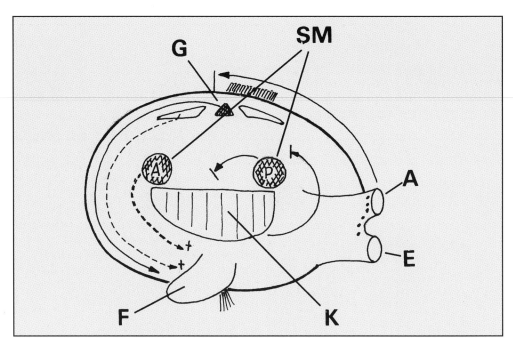

Diagram of a Tridacna larva before metamorphosis. The ligamentous joints are on top (G), the two breathing syphons (E+A) are located on the side and the gills (K) in the central part of the animal. The shells are closed with the help of two adductor muscles (SM). The foot (F) as well as the byssal apparatus is on the bottom side of the larva.
Drawing by Melchior Buelo according to a copy by Dr. H.-H. Janssen

the animal. The drawing below shows that the *Tridacna* larva is not different from any other mussel before its metamorphosis. The shells are connected on its upper side by a ligament and thus open downwards. The animal sticks its foot and byssal apparatus through this aperture to locate and attach itself to the ground. The breathing syphons, one for intake, one for outlet, stick out of the shell on one side. The gills are located in the centre of the mussel. The two shell halves are connected by two different adductor muscles: the front one called musculus anterior, and the rear one called musculus posterior. Because of these two adductor muscles the larva is called

dimyaric which stands for " with two muscles".

The metamorphosis which takes place in the larva is obviously one of the most interesting phases in its life. It aims at providing the symbiotic algae with optimal living conditions. Our larva needs a large body surface to house as many symbiotic algae as possible. On the other hand, the surface has to be flexible to be able to retract into the shells if danger approaches. In the beginning of the metamorphosis, the dinoflagellates of the species Symbiodinium microadriaticum Freudenthal, which have been filtered out of the ocean by the clam larva, are

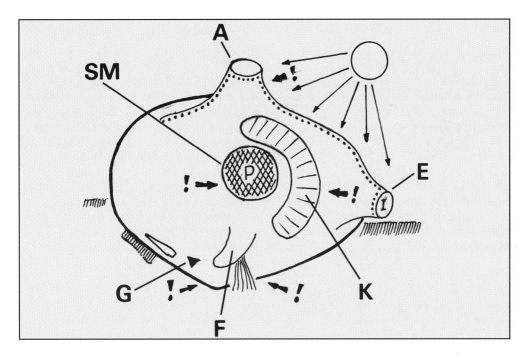

Tridacna larva after metamorphosis. The ligamentous joint of the shells (G) has moved downwards and the excurrent syphon (A) is now at the top. The shells are closed by a single centrally located adductor muscle.

Drawing by Melchior Buelo according to a copy by Dr. H.-H. Janssen

deposited in the tissue between the two breathing syphons. In the following stage, this part of tissue prolongs, and the excurrent syphon moves towards the upper side of the animal, while the incurrent syphon remains in its place. At the same time the shape of the shell is changing substantially: The ligamentous connection of the halves, which is originally situated on top, moves in one direction around the animal to the bottom, thus enabling the shells to open on the upper side. In doing so, the larva has provided room for a vast, extensive mantle and the tissue between the two syphons can proliferate massively.

Juvenile Tridacna derasa at the age of a few months. Shell length is about 5 to 10 mm.

At the bottom side, however, room is in short supply, as the mussel still needs to locate its foot and byssal apparatus on the ground. This is why it develops an additional aperture, the byssal orifice. Due to this bilateral opening directly next to the ligament of the shells, the animal is able to attach itself to the ground.

The inside of the larva also undergoes dramatic changes. The front adductor muscle disappears and the rear muscle moves into the centre. As a consequence the animal is now no longer called dimyaric, but monomyaric, which means "with one muscle". The gills which have so far been located at this place in the centre of the animal, are at the end of this development, curved like a "banana" around the adductor muscle.

This fascinating metamorphosis which turns our little larva into a clam within three to six weeks after fertilization, is due to differences in growth. While in some body parts growth is amplified, in others it is diminished, so that as a consequence, the shape is altered. The result of this alteration is a mussel which can cover its complete light exposed upper surface with a leaf like organ. This organ is used to propagate unicellular plants that serve to feed the clam with its metabolic products.

Our little clam has now its own symbiotic algae and settled in a place appropriate for its needs. Its shell size is roughly one millimeter. The first and extremely risky stage of its life as part of the plankton is over, but also the sessile life on the litoral zone, poses dangers to the animal. It is so tiny, it is threatened mainly by algae eating organisms, which eat it by mistake when foraging on the algal cover. When it grows bigger, to about three or four millimeters in length, it becomes more conspicuous and develops into interesting prey for small labrids and other mollusc eaters. Growth is enormous at this stage and after a few months, the clam has - depending on its species - reached 20 to 40 millimeters in length.

Still the young clam is threatened by predators. With a shell size of two centimeters it is big enough to be recognized by grazing surgeon fishes and other algae eaters, but its shells are still weak and fragile. The sharp scales

Juvenile Tridacna crocea at the age of a few months. (A German coin of about 2 cm in diameter serves as a reference)

eventually reach the less risky juvenile stage, which is when the clam reaches a shell size of 5,10 or 20 centimeters depending on the respective species. The strength and stability of the shell as well as the attachment to the ground by the byssal filaments provide effective protection. Some species have evolved special protection mechanisms, like for example digging the shells into the ground in between the rocks and broken coral or boring the shells into the solid limestone. In both cases only the mantle of the animal is visible which can be retracted rapidly in case of danger. The predator is then left with two closed shell rims armed with teeth.

which stick out of the shell at right angles provide protection to a certain extent, yet larger predators specialized on crustaceans and mussels do not worry about these scales. Many of the small clams therefore lose their life at this stage and just a few survive and

A few years go by before the few surviving clams reach male sexual maturity and can participate in mating. Some of them may one day become fully mature, and also produce eggs, which is the start for a new reproductive cycle.

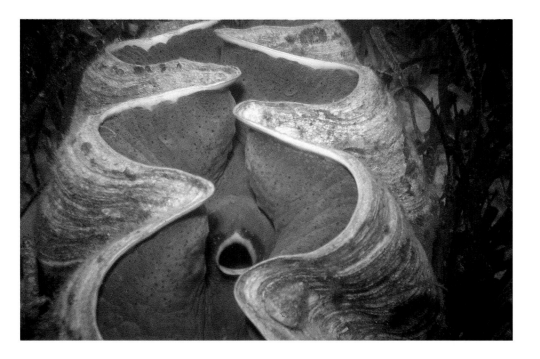

4 Anatomy

Introduction

Like all mussels, the clams of the family *Tridacnidae* have an outer skeleton consisting of two shells. In contrast to most other mussel species the representatives of the *Tridacnidae* have changed their anatomy and physiology dramatically to adapt to the presence of symbiotic algae in their body tissues. In the preceding chapter these changes have been described in detail. This development is an adaptation of the host to the symbiotic algae which is termed "symbiont induced co-evolution" in the biological sciences.

The *Tridacna* clams which belong to the *Cardiaceae* (heart shaped mussels) are very different from the typical representatives of the family. Although the presence of symbiotic algae in the mantle is rare in molluscs, numerous other members of the family have a highly developed symbiotic way of living with unicellular algae and adapted to it anatomically and physiologically. Interestingly science has noticed this ability relatively late (Kawaguti, 1959, 1968, 1983; Janssen, 1988, 1989) although some of the species are known since long. The mussel *Corculum cardissa,* for example, which also lives in symbiosis with algae,

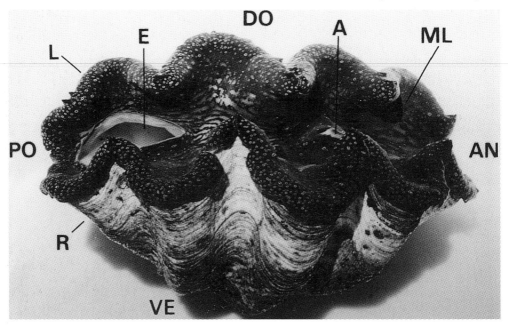

Overall view of a giant clam (Tridacna gigas, shell length about 35 cm). E = incurrent syphon, A = excurrent syphon, ML = mantle lobe (syphonal mantle), DO = dorsal (back side), VE = ventral (belly side), AN = anterior (front), PO = posterior (back), L = left, R = right.

Photo: Dr. John Norton

has been described by Linné as early as 1758.

In the meantime some 20 species of symbiotic living mussels have been scientifically described. However, some of these descriptions are unpublished, so that the species and their way of living is unknown among today's sea water aquarists. In addition there is another much larger group of mussels which has developed a symbiotic way of living with bacteria. The appendix at the end of the book reports on these species in more detail.

The mussels of the family *Tridacnidae* are provided with nutrients from the photosynthesis of their algae to complete their diet (Fitt, 1988). Although they still have gills which are common in mussels and which filter planktonic food from the water, they have achieved a lot of independence from the nutrient densities of their habitat by endosymbiosis. Molluscs living in symbiosis with algae were capable of inhabiting reef areas where food is much less abundant as compared to their relatives that depend entirely on filtration. The less suspended food in the water, the less turbid and the cleaner it is; thus the more light can penetrate.

Yonge (1975) assumed that it was the endosymbiosis and the accompanying better nutritional state, which enabled the clams to overcome the usual mussel sizes by far. If this was so, the largest species among the giant clams, T. gigas, should have developed the most intensive use of endosymbiosis.

The endosymbiosis with the algal species *Symbiodinium microadriaticum*, is not unique amongst the other marine

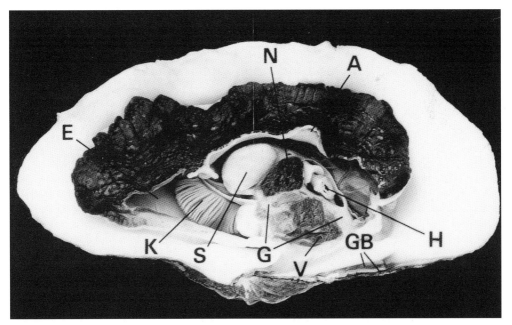

Mid-longitudinal section (Tridacna gigas, shell length about 35 cm). S = adductor muscle, N = kidney, H = heart, K = gills, V = digestive organ, G = gonads, GB = ligaments.

Photo: Dr. John Norton

invertebrates. Most of the corals of the litoral zone have taken up this way of supplementary nourishment. Stony and soft corals of a variety of species house these algae in their body tissues. However, there is a significant difference between these corals and the clams: in the corals, the microscopic algae are located intracellularly, that is, within the cells, whereas in the Tridacna clams they are extracellular, that is, in between the cells in a special channel system of the syphonal mantle.

The organs of the clam are enclosed in a coat that is strongly attached to the inner side of the shell. The clams have two pairs of gills, used for gas metabolism and for filtering suspended food. They have a mouth, an oesophagus, a stomach and intestines. Two renal lobes serve to detoxicate the body and the heart pumps lymphatic fluid through a simple circulatory system. Male and female gonads are present. A foot and the byssal apparatus help the animal to attach to the ground. A simple nerve system runs through the body tissues of the clam.

The Individual Organs

The Shell

Mussels have no inner skeleton, but an outer one: the shell. The two shell halves of the giant clams of the family *Tridacnidae* are like convex chalky plates, connected by a joint at the bottom. Four to six vertical folds emerging from the bottom of the mussel near the joint run across the entire shell to the upper rim and give it its more or less corrugated surface. This corrugation protrudes into the rim which gives it an undulating appearance, resembling almost teeth. In most species the teeth of opposite shell halves fit perfectly into each other and thus allow the clam to close the shells completely. Only in the adult individuals of *T. gigas,* is this closing incomplete, and part of the mantle is always visible, even when the shells are closed.

The surface of the shells is grey-white and coarse. Besides the vertical folds, the shell also comprises horizontal rib like lines, which are kind of a growth marker, similar to the growth rings of a tree. The distance between two of these lines allows conclusions on the growth of the shell within one growing period. The duration of these periods and the speed of growth differ between species substantially. Some species develop sharp scaly protractions alongside those growth lines which always emerge from the vertical folds and stick out, at almost right-angled into the water (*T. squamosa,*

T. maxima, T. rosewateri). Above the scales chalky pouches develop, commonly used by various organisms for shelter. A rich fauna is found here: small mussels which are strongly attached to the host mussel, sessile animals like sponges, tube worms and other filtrating organisms and even stony corals. Occasionally small crabs emerge, which are inconspicuously coloured and therefore can only be recognized if you take a close look.

Not all *Tridacna* species develop these scales. Some species have smooth shells and only the fine horizontal lines are visible. One species develops short ribs which are densely packed and make the shell appear like a file. In fact, this clam uses the scales like a file to bore itself into the soft limestone (see description of species T. crocea).

At the bottom of the shells is the byssal orifice. This oval opening has a typical shape in most species and therefore provides a good criterion for species identification. At the joint side of the orifice, there is a series of teeth-like lamellae, usually four to seven, which are densely packed closer to the joint.

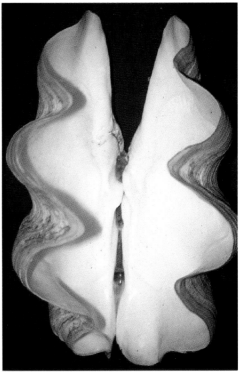

A pair of shells from above (Tridacna gigas, shell length about 35 cm)

The shell of a giant clam from the outside. The vertical folds and horizontal lines are clearly visible. (Tridacna gigas, shell length about 35 cm).

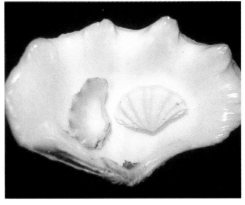

Occasionally an unusual shell colouration with a more yellow or orange tone occurs. The photo shows Tridacna maxima (large), Tridacna crocea (small left) and Tridacna gigas (small right).

The shell halves are connected mechanically by a joint at the bottom. It consists of four protrusions, each pointing towards a pit on the other shell half. If you look at the inner side of the shell and focus on the lower rim, you will discover a so-called cardinal tooth and a lateral tooth on each shell half. The right cardinal tooth, which is longer and situated very close to the byssal orifice, is the most important part of the joint which never fully retracts from its pit on the left shell half. Even when the shell is maximally opened, at least its tip remains in the pit. The left cardinal tooth moves over the right one and terminates in a pit in the right half of the shell when closed.

The lateral tooth is located laterally (on the side). This tooth reaches its pit only when the shell is closed and provides a second anchor point to prevent the shells from shearing stress. When the shell is open it retracts completely from the pit. This lateral tooth is also present in both shell halves.

The four teeth, and the respective pits allow the joint to open only in one direction like a hinge. These anatomical structures are however not interconnected, but just inserted into each other and therefore do not support the connection of the two shell halves. This connection is provided by the inner and the outer ligament. They are situated next to the four teeth, start directly after the byssal orifice and end usually half way to the end of the shell. The outer ligament is longer and significantly thinner than the inner one. Between the two are chalky enclosures which serve, according to my opinion, to prevent boring parasites from penetrating into the mussel through the ligaments.

The inner side of the shells is white. At

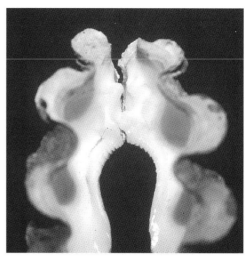

The joint region with the cardinal and lateral teeth (Tridacna maxima, shell length about 11 cm).

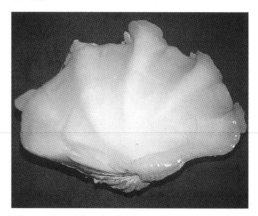

The inside of the shell. The pallial line and the point of attachement of the adductor muscle, as well as the two dark ligaments can be seen (T. squamosa, shell length about 25 cm).

a closer look, you can see that the central part of the area is rather blunt and mat, whereas the more distal parts are shiny like porcelain. The border between the two parts is a fine line called the pallial line. It starts down at the joint and runs across the whole shell in an oval line to end again next to the joint. The retraction muscle of the mantle inserts along this line. Below the line within the blunt area, the soft parts of the mussel are connected with the shell. The shiny part of the shell above the line is the area where the mantle protracts and retracts. Above the pallial line a round scar is visible. The adductor muscle which connects the shell halves and which can close the shell with great vigour is fastened here.

The very chalk-rich shell grows at the upper rim, below the overhanging mantle. The processes for shell growth are, however, not yet fully understood. In a manner so far unclear, the clams take up calcium ions from the sea water. With the help of carbonic acid metabolized by the animal, the calcium turns into calcium bicarbonate and in a further step into calcium carbonate. Part of this calcium carbonate is dissolved (chalk-carbonic acid- equilibrium). The symbiotic algae of the clam consume carbon dioxide and thus part of the dissolved chalk is precipitated; The precipitated chalk is now available for the construction of the shell.

The Syphonal Mantle

The mantle covering the intestines of the clam consists of two parts: a muscular pigmented syphonal mantle and a thin transparent lateral mantle. The mantle has three apertures: the incurrent and excurrrent orifice at the upper side and the byssal orifice at the bottom.

The syphonal mantle covers the surface between the upper rims of the shells and overlaps the shell rims in almost all species. This mantle is attached with numerous muscles to the inner side of the shell at the pallial line.

The lateral mantle comprises thin, unpigmented tissue covering the inner surface of the shell from the foot up to the pallial line. This lateral mantle is attached to the shell over its entire surface. Within the byssal orifice where the flexible foot of the mussel can be retracted, the collar-like pair of mucus glands, the pallial glands are situated. Its exact location can be neglected here - like a few other anatomical details - in favour of a more clear description of the anatomy of these animals.

When taking a closer look at the surface of the mantle, you can see a row of thin elevations on its rim, the so-called iridophores. These are small eye organs which comprise a simple lens. However, these very primitive eyes are not good enough to give the mussel a picture of its environment. It cannot recognize objects, but it can measure the amount of incoming light. This helps the animal to recognize day and night, and induces the protraction of the

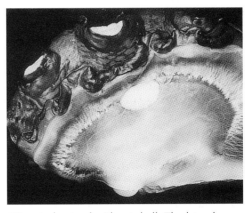

Dissected animal without shell. The lateral mantle which is attached to the shell in the live animal and the retracting muscles of the syphonal mantle which are connected to the shell at the pallial line are visible. (Tridacna gigas, shell length 35 cm). Photo: Dr. John Norton

mantle towards the sunlight. In addition the iridophores tell the nervous system, if a shadow falls on the body, which indicates that a predator is approaching. The very primitive nervous system of the clams must not be misunderstood; the mussel does not actually fear a predator, as it does not know emotions or experiences. For that a much more complex nervous system would be necessary. But the mussels have undergone a selection process in evolution, where all the individuals which were not quick enough in closing their shells were eaten by predators. Only the animals with a very quick reaction could survive and reproduce and thus the ability spread in the population. The clams we know today are able to react very quickly with a reflex which can be compared to our coughing reflex.

The lenses of the iridophores have a very interesting side effect. They bundle the light and direct it into the lower layers of the mantle. This is the reason why the symbiotic algae are concentrating around the iridophores which is indicated by a dark ring around the base of the little elevations.

Under favourable conditions, the symbiotic algae proliferate very quickly in the mantle. Therefore the animal needs a mechanism to get rid of the surplus algae. The mussel solves this problem by "harvesting" regularly in its "vegetable garden". Older algae are sorted out and are shed. If overpopulation threatens the mantle, even young functional algae are attacked and disposed. A kind of a "health police" is responsible for that: macrophages moving around in the tissue in an amoebic manner destroy the surplus algae. This is necessary, for example, when the amount of light is increased during the summer season by

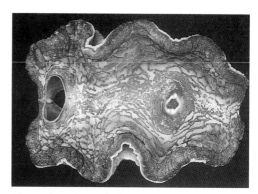

The syphonal mantle from above of a dissected clam without shells. (Tridacna gigas, shell length 35 cm). Photo: Dr. John Norton

more intensive sunlight. Each and every microscopic alga assimilates more in more intense light and also produces more oxygen. To prevent oxygen intoxication, the clam has to reduce the amount of its symbionts. This is visible externally by a lighter brown colouration of the mantle.

But the mussel must also take measures in the opposite case. For example, if the light intensity is reduced by thick clouds in the tropical rainy season, the algae can assimilate less. The amount of metabolic products is therefore reduced per individual, and the mussel has to breed more algae to get the same amount of oxygen and nutrients and to get rid of its own toxic metabolic products, which in turn are the basis of the algae's nutrition.

In case of low light intensity, the clam will tolerate a high density of algae on its mantle and will not reduce its vegetarian guests. This leads to a much darker brown colouration of the mantle. If, on the other hand, the algae die due to a lack of light, the mantle is

Syphonal mantle of a Tridacna gigas with typical colouration. The blue skirted iridophores at the rim of the mantle are clearly visible.

Differently coloured clams in an aquarium.

Photo: Enzmann

significantly lighter in colouration. In the chapter "Clam Diseases" these mechanisms and their pathology is explained in greater detail. In case of high light intensity, the mussel realizes the increased amount of assimilation products in its tissue and not, according to my opinion, the increased amount of light, as has been suggested by some authors, and it reacts with a reduction of the algae as described above.

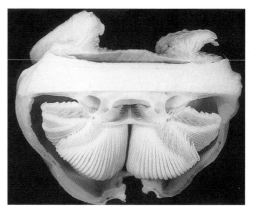

Three transverse sections in different layers of a dissected animal. The photos show the incurrent chamber and the gills (Tridacna gigas, shell length about 20 cm).   Photos: Dr. John Norton

The Water Chambers

The lobes of the mantle form the external boundary of the water chambers, which are enclosed in the syphonal mantle and bordered by the lateral mantle inside the shell. This chamber system can be divided into two major chambers: the incurrent chamber and the excurrent chamber. The incurrent chamber is located directly behind the incurrent syphon and comprises the gills. When the water passes through the gills, it reaches the excurrent chamber which is substantially larger and stretches along

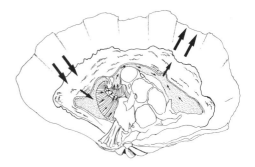

Water circulation through the chambers. Drawing: Melchior Buelo according to a sketch of Dr. John Norton

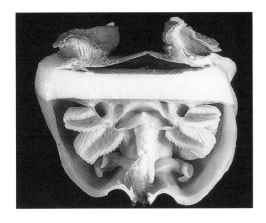
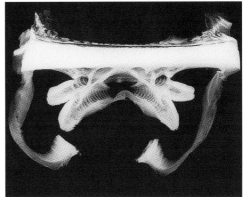

almost the entire length of the mussel's body. The water leaves the excurrent chamber via the excurrent syphon situated in the middle of the syphonal lobe.

The main chambers are not only separated by the organs between them, but more importantly by a thin transparent skin, the interchamber membrane. The animal can fill either one of the two chambers, by opening of the shells - which is usually the incurrent chamber - or even both chambers which can become necessary if a small particle got stuck in the chamber system and has to be "coughed

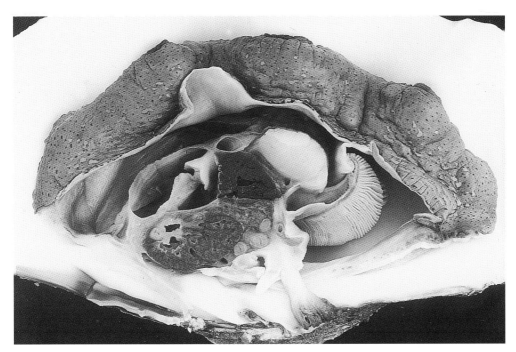

Mid-longitudinal section through a dissected clam. The incurrent chamber in front of the gills (right) and the excurrent chamber behind the gills (left) which extends from the gills to the end of the syphonal mantle.
Photos: Dr. John Norton

63

out" by a vigorous contraction. If only the incurrent chamber is to be filled, the animal closes the excurrent syphon and opens its shells. The middle part of the syphonal mantle is then slightly retracted above the excurrent chamber.

A similar mechanism is responsible when the excurrent chamber has to be emptied. The mussel closes its incurrent syphon and contracts in a usually jerky manner. The water leaves the chamber as a strong jet from the excurrent syphon. This mechanism is sometimes used by large animals to reject attackers or to scare away molesters.

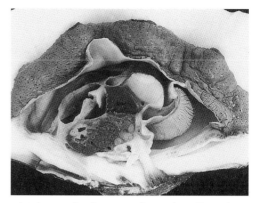

This longitudinal section shows the gills in the incurrent chamber.   Photo: Dr. John Norton

The Gills

The gills consist of a pair of laminated cords located in the incurrent chamber. They are oriented vertically so that the incurrent water hits their surface. The lamellae with which the molluscs filter the water are arranged bilaterally on all four cords and they are densely packed together, so that the relatively small gills provide a very large surface.

In giant clams these organs are, in contrast to the gills of a fish, not only a means of gas exchange, but they also filter the water for suspended food, as is

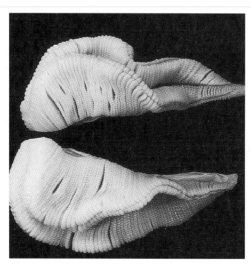

The gills: In this isolated preparation the nutrient channels are easily visible where the filtered suspended food is transported to the mouth.
Photo: Dr. John Norton

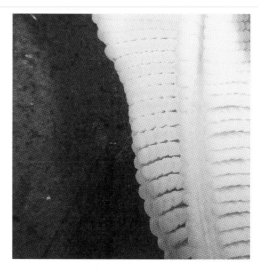

Enlargement of the gill lamellae: The corrugated gills provide a large surface for gas exchange and filtration.
Photo: Dr. John Norton

the case in all mussels. The food particles are transported towards the mouth via a system of several food channels and then pass the digestive tract. One of these food channels can be seen when looking into the incurrent chamber: it is the longitudinal cleft running across each of the four laminated cords. The animals are, however, very choosy in their diet. They only accept the extremely fine food particles, the so-called "nano-plankton". Any larger plankton particles and other dirt particles are pressed into flakes and expelled through the incurrent syphon.

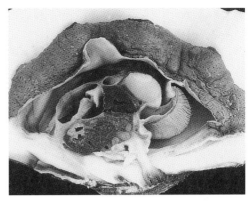

This longitudinal section shows the adductor muscle which is situated at the centre of the shell directly below the pallial line. It consists of two parts: the white muscle (above) and the hyaline muscle (below).

Photo: Dr. John Norton

The Adductor Muscle

In contrast to other mussels which have two adductor muscles, the clams of the family *Tridacnidae* are provided with just one. This muscle is a white, cylindrical organ stretching from one shell valve to the other. Its point of attachment is located in the middle of the two shell halves at the upper rim of the pallial line. It consists of two parts which cannot be told apart when looking at the intact muscle. Only when a cross section is made through the muscle, do the two parts become apparent. The upper cord which is called white muscle is much thinner than the lower; and it serves to close the shell on a long-term base. The lower cord, called the hyaline muscle is the rapidly reacting part of the adductor. It serves to act immediately and quickly, when the clam is disturbed, endangered or when it expels an alien particle.

In the centre of the upper side of the adductor muscle, the anal papilla is visible comprising the outlet of the digestive tract which is connected to the muscle. The faeces in the form of a thin brown thread is excreted into the water through the anal papilla via the excurrent syphon. At the lower side of the muscle a large and important bundle of ganglia is situated responsible for the nervous control of the intestines.

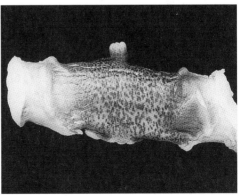

Isolated preparation of the adductor muscle. The anal papilla is clearly visible (on top). Interestingly also the adductor muscle comprises the symbiont channel system in its light exposed parts. The dark colouration is therefore due to symbiotic algae.

Photo: Dr. John Norton

The Kidneys

Clams have two kidneys to cleanse the blood lymph of the circulatory system from toxic substances. These two organs are located on both sides below the adductor muscle almost precisely in the centre of the animal. When one shell is removed during dissection, the kidneys are clearly visible due to their maroon red colouration and their round shape. The kidneys which are connected to the circulatory system, are continuously provided with blood lymph which they detoxify. The outlet of the kidneys is on the upper side and directed slightly outwards towards the shell. The secretions of the kidneys are flushed into the excurrent chamber and excreted through the excurrent syphon.

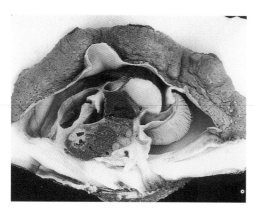

Mid-longitudinal section through the animal. The kidneys are located below the excurrent syphon, almost at the centre of the clam, next to the adductor muscle. Further towards the front, the pericard is situated, which is opened in the preparation above, so that the the heart and its valves are visible. Photo: Dr. John Norton

The Heart

The heart of the giant clams is situated right next to the two kidneys, almost in a direct line below the excurrent syphon. If the mantle is opened from its anterior end towards the excurrent syphon and its parts tipped to the side, the pericardium which encloses and protects the heart is visible. The pericardium is a chamber slightly larger in volume than the kidney. In this chamber, which I shall call pericard chamber, the heart is placed with its pumping muscle. If the pericardium is opened with a central cut, the heart can be seen.

This surprisingly small and simple heart consists of several chambers and leads into the two main arteries. Each connection is provided with valves to prevent the blood lymph from running back.

The Circulatory System

The Arteries

The circulatory system runs from the heart in two main arteries through the tissue of the mussel. The front aorta (Aorta anterior) runs from the heart downwards and supplies the digestive tissue and the gonads with their eggs and sperm via the visceral arteries (viscera = internal organs). The endings of this aorta supply the byssal gland and the foot of the animal with oxygen and nutritive substances.

The back aorta (Aorta posterior) runs from the heart upwards and supplies the adductor muscle and the syphonal mantle. The artery splits into two branches, a left and a right one to reach both sides of the mantle. It branches into two further arteries to supply the front and the back half of the mantle and then splits up into a fine capillary network.

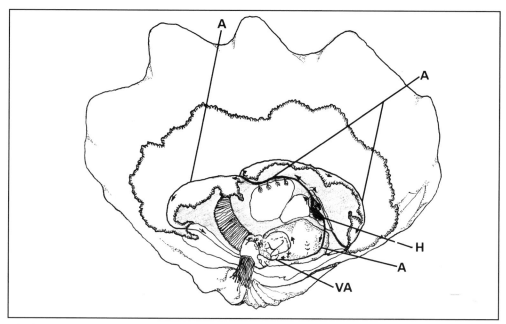

The location of the arteries drawn from a longitudinal section. H = heart, A = artery, VA = visceral arteries.
                                   Drawing: Henri Rivero according to a sketch of Dr. John Norton

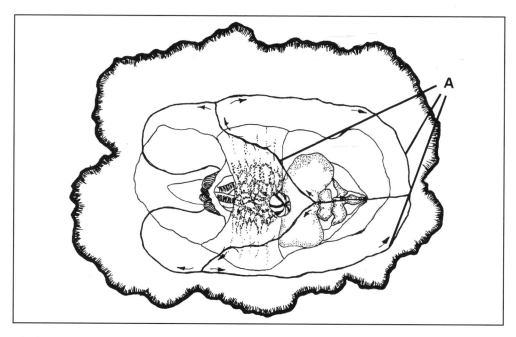

The location of the arteries as seen from above. The blood lymph runs in the direction of the arrow. A = arteries.
                                   Drawing: Henri Rivero according to a sketch of Dr. John Norton

The Veins

While the arterial system transports the oxygen rich blood lymph into all parts of the body and supplies all organs with nutrition, the venous system provides the back transport of the blood lymph. The system is comprised of clefts in the mantle, the digestive organs, the kidneys and the adductor muscle which transport the blood lymph to larger veins which end in the large pallial vein. This very large vein which runs between the adductor muscle and the kidneys eventually ends at the gills. In the gills the blood lymph exchanges toxic substances for new oxygen. The first part of the venous system ends here.

Before the blood lymph can reach the second part of the venous system, it has to pass through the gills. Here, the biophysical processes take place which free the loaded blood lymph from toxic substances and provide it with fresh oxygen. Each single gill lamella encloses a number of thin veins. Both parts of the venous system, the first as well as the second part have thin veins in the gill lamellae and thus secure the exchange of the vital blood lymph.

The second part of the venous system transports the blood lymph back to the heart, where it is pumped again through the organism.

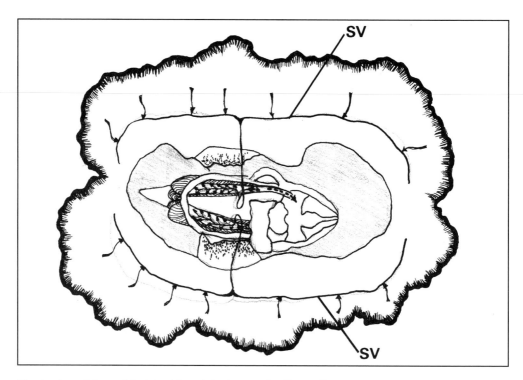

The backward flow of the blood lymph in the mantle. SV = collective vein.
Drawing: Henri Rivero according to a sketch of Dr. John Norton

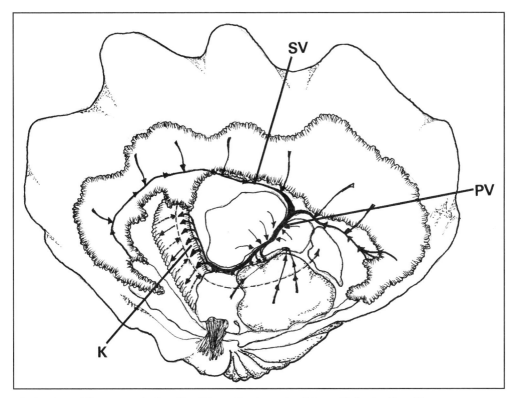

The backward flow towards the gills. SV = collective vein, PV = pallial vein, K = gills.
Drawing: Henri Rivero according to a sketch of Dr. John Norton

The Digestive Tract

The common view of many sea water aquarists that the giant clams live solely from the metabolic products of their symbiotic algae does not hold true if you take a closer look at their anatomy. Although it is correct that the endosymbiosis is a vital base of the clams nutrition, also filtration forms an important source of food for these molluscs. They have a digestive apparatus comprising of several singular organs in a muscular mantle which is located directly under the pericardium.

The digestive tract consists of a system of nutritional channels, where the suspended food from the water is transported from the gills towards the mouth. Correctly speaking, the digestive system entails the gills themselves, as the gills have already these kinds of channels where the nutritional particles run downwards. From the channels the food reaches the mouth which is located directly above the hinge of the two shell halves and enters the oesophagus. The food is stored in the stomach, mixed and broken down by digestive secretions from special glands. During the following passage through the intestines, important nutritional substances are taken out of the food and secured for the organism. The rest is transported to the anus, which is situated in the excurrent chamber directly below the excurrent syphon. The light water current which permanently runs through this syphon carries the excretions of the clam into the open water easily visible as a thin brown thread emerging from the syphon.

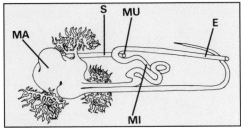

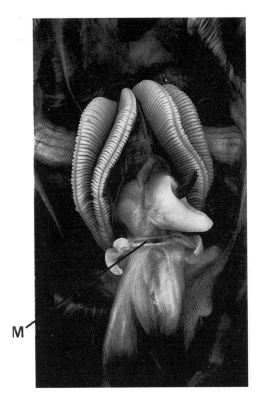

▲ Model of the digestive system. MU = mouth, S = oesophagus, MA = stomach, MI = middle part of the intestines, E = end part of the intestines.
Drawing: Melchior Buelo according to a sketch of Dr. John Norton

◀ The mouth at the bottom of the animal from behind the byssal gland. M = mouth.
Photo: Dr. John Norton

Mid-longitudinal section through the animal: Below the kidneys and the pericard, the digestive organ is visible, where the entire digestive tract with all its glands is placed. On both frontal sides of this digestive apparatus, the light coloured tissue of the gonads can be seen. Here the sperms and egg-cells are produced. G = gonads, V = digestive organ.
Photo: Dr. John Norton

▼

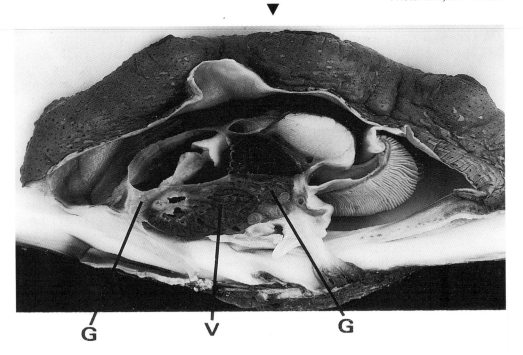

The Gonads

As has been mentioned above, clams are hermaphrodites. The word comprises the name of the ancient greek messenger "Hermes" and the goddess of beauty "Aphrodite". An organism called hermaphrodite has both genders - it is male and female in one.

In contrast to some organisms which are male and female at different times of their lives - for example certain shrimps, which are male when young and female when fully grown - the clams are both at the same time, which is called simultaneous hermaphroditism in science. This is at least true for the adult specimens which are fully grown whereas, when young, only the male gonads are active.

The reproductive system of the clams consist of clusters of tissue around the digestive tract. Particularly dense clusters of these cells are mainly found in two places of the body. One is at the beginning of the digestive tract below the kidneys, roughly in the centre of the animal, the other is at the anterior end of the digestive system. In juveniles, both clusters comprise immature gonadal cells which are connected to a wide net of ducts. These ducts lead to larger common ducts and eventually to the gonadal orifice.

With increasing age and size of the animal, the gonadal cells develop into male testes and later, when the animal is close to maturity, into female ovaries. The products of these organs, the sperms and egg cells (oocytes) are released into the excurrent chamber during reproduction and pumped with the water current through the excurrent syphon. As has been mentioned earlier, the simultaneous release of both sperm cells and egg cells is not possible and a temporal delay of 20 to 30 minutes is necessary. This prevents the mussel from self-fertilization, which can cause genetically based handicaps in a high proportion of its progeny after a number of generations. Experiments with artificial self-fertilization have shown that the survival rate in that kind of offspring is significantly lower than in offspring with genetically non identical parents (Erwinia Solis-Duran, 1994, pers. comm.).

The Foot and Byssal Apparatus

The byssal apparatus is an organ known in most mussels. This gland, or correctly speaking, the byssal organ and the gland are particularly important for the three smaller species, *T. crocea, T. maxima* and *T. squamosa,* whereas the larger species attach to the substrate only as long as they are juveniles. Once they are fully grown, their weight holds them on the ground. This is also true, to some extent, for the larger specimens of the above mentioned smaller species. I have seen fully grown *T. squamosa* which were not showing any sign of attaching themselves to the surface with their byssal organ over a long period of time. The smaller individuals, however, spend their entire lives strongly attached to the ground.

To do so, they use the byssal threads, which are extremely strong and which stick to the chalky substrate almost like glue. The threads are produced in the byssal gland in the form of a fluid, which hardens when in contact with water.

The byssus organ with the byssal gland is connected to four muscles similar to suspension in gimbals. Strictly speaking there are just two muscles

Isolated byssal gland with the two upper retracting muscles. Photo: Dr. John Norton

A T. crocea searching for a new settling substrate at night. The white foot is prominent.

which come in pairs with a right and a left branch respectively. One of the muscle pairs holds the byssal organ on its upper side. The endings of the two muscle branches amre connected to the inner side of the shells directly below the pallial line and next to the large adductor muscle. The other pair of muscles which is considerably shorter, is connected to the backside of the byssal organ and stretches to the bottom of the shells where it ends on both shell halves close to the cardinal teeth.

The byssal organ is relatively mobile within the shells. The mussel can lower it to attach itself with new threads to the substrate, but it can also retract it into the inside, which can be released by touching the soft tissue at the byssal orifice. The suspension on four muscles is, however, a very weak point as this mechanism is not resistent to strong tractive forces. This mechanism protects the animal against strong currents and it also ensures that the bottom of the shells are attached to the substrate as tightly as possible, avoiding any gap of which predators would take advantage. If one tries to tear off an animal from its settling substrate with vigour, the muscular attachment apparatus or the byssal organ itself can easily be injured. Such an animal is doomed and will never recover.

The Nervous System

Like all mussels the giant clams do not have a brain. The control of the nervous system is provided by four ganglia. These ganglia comprise of knot-like bundles of nerves with a particularly high number of neurons (nerve cells). From here the nerve fibres emerge which run through the entire body tissue of the animal, each fibre originating in a neuron of the ganglia. Besides the many short projections which interconnect the neurons within the ganglia to exchange information, the neuron has one very long projection, which is the actual nerve fibre. This fibre transmits the information to all distant body parts. Due to the electrical impulses transmitted by these fibres, body movements or blood vessel contractions are released and sensitive information received, the nutritional status of certain body tissues controlled, and the activity of particular glands stimulated. The further the fibres intrude into the tissue, the more they are branched forming

finer and finer fibres which eventually cover the whole body like a dense network.

As has been mentioned above, the clams have four ganglia, where the neurons are interconnected, to exchange information. One of these ganglia is the foot-ganglion which controls the foot and byssal apparatus. It is situated directly on the foot itself, next to the digestive organ.

Very close to this ganglion another ganglion is placed which comprises a pair. There are actually two ganglia which are interconnected by a direct nervous line. They are called cerebral ganglia and are located at the bottom of the digestive apparatus.

The largest of the ganglia is the visceral ganglion which controls the majority of the organs as well as the syphonal mantle. It is linked to the other ganglia. The pallial nerves emerge from the visceral ganglion which control the

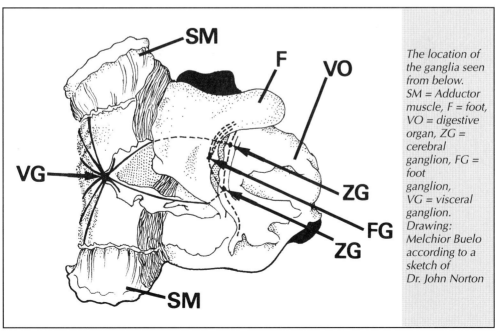

The location of the ganglia seen from below. SM = Adductor muscle, F = foot, VO = digestive organ, ZG = cerebral ganglion, FG = foot ganglion, VG = visceral ganglion. Drawing: Melchior Buelo according to a sketch of Dr. John Norton

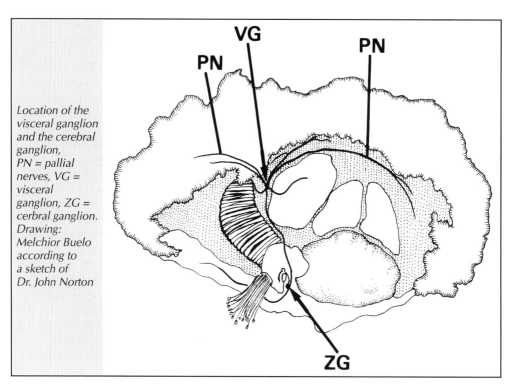

Location of the visceral ganglion and the cerebral ganglion, PN = pallial nerves, VG = visceral ganglion, ZG = cerbral ganglion. Drawing: Melchior Buelo according to a sketch of Dr. John Norton

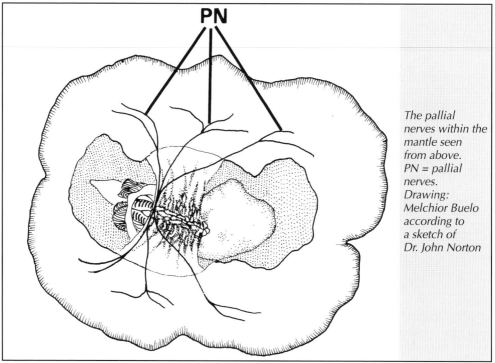

The pallial nerves within the mantle seen from above. PN = pallial nerves. Drawing: Melchior Buelo according to a sketch of Dr. John Norton

mantle - for instance, they control the retraction muscles which retract the mantle towards the pallial line when the shells are closed. The visceral ganglion is located close to the bottom of the adductor muscle and directed towards the gills.

The Symbiont Channel System

The symbiont channel system is a specialty of the clams which separates them from other molluscs of the class Bivalvia. This system has been discovered only very recently: In May 1992 it was published by scientists from Queensland and Georgia (J.Norton, M. Shepherd, H. Long and W. Fitt, 1992).

The idea, that the symbionts of the clams live in a channel system especially developed for them is not new, as it has been proposed already in the middle of this century (K. Mansour, 1946). Mansour even partly described the channel system, and reported that it would run from the stomach into the mantle and would contain the symbiotic algae. The fierce response by the great researcher Sir Maurice Yonge, who vehemently denied the existence of such a channel system (C. M. Yonge, 1953 a+b) must have intimidated Mansour, as he did not continue to examine the channel system. It was commonly assumed that the symbionts of the giant clams live in the blood vessels of the syphonal mantle. In 1992 it was shown by Norton and his colleagues that the symbiotic algae of the giant clams actually live in a closed channel system with no connection to the blood vessels.

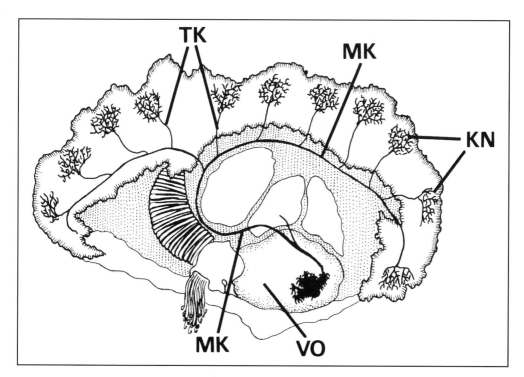

Schematic longitudinal section to demonstrate the symbiont channel system. VO = digestive organ, MK = mother channel, TK = daughter channels, KN = channel network.
Drawing: Melchior Buelo according to a sketch of Dr. John Norton

Curiously the channel system proceeds from the stomach from where in juveniles the first symbiotic algae reach the channel system. It is not clear, how these algae which are filtered out of the water can escape digestion by the stomach. The unicellular algae reach the mantle via a direct connection between the stomach and the channel system comprising of an orifice (W. K. Fitt and R. K. Trench, 1981) visible in the animal at an age of only a few weeks (P.S. Lee, 1990).

On the other hand, old or superfluous algae reach the digestive tract the same way, and are excreted without being digested (Norton, 1992). The protective mechanism that saves the algae from being broken down by the digestive enzymes is also not yet clarified. I assume that the protective mechanism is found in the algae, not in the mussel. The algae could have evolved a specific resistance against the particular secretions of their hosts. The primary benefit of not being killed is obviously on the side of the algae.

From the stomach a singular channel runs upwards, and splits up into two branches while still in the digestive tissue. These two channels perforate the muscular covering of the digestive organ and wind around the adductor muscle upwards to the syphonal mantle. In the vicinity of the pallial line, the two branches split again, so that a front and a back channel is formed. Thus there are four branches of channels which supply a quarter of the overall mantle surface respectively. These branches run

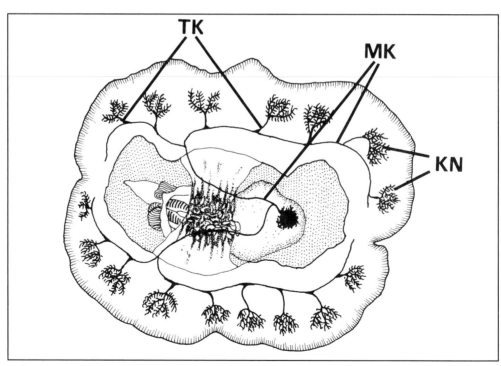

The symbiont channel system within the mantle seen from above. MK = mother channel, TK = daughter channels, KN = channel network. Drawing: Melchior Buelo according to a sketch of Dr. John Norton

through the entire mantle parallel to the pallial blood vessels and nerves. A bundle of four ducts is found here: an artery, a vein, a pallial nerve and a symbiont channel.

In the course of their progression, a number of smaller channels, "daughter channels" emerge from the major channels, which I would like to term "mother channels". The daughter channels progress rectangularly into the tissue of the mantle and branch off to form a fine network of very thin channels. There, the symbiotic algae are found which provide the clams with food for their metabolism. They are imbedded in the mantle itself, solely within their channel system and not in the cells of the tissue, as is the case in corals.

The daughter channels which emerge from the mother channel do not only progress into the mantle of the animal, but also into tissue surfaces below the mantle which are sufficiently exposed to light. For instance, often light can penetrate into the inside of the mussel through the incurrent syphon. To make use of this light for the photosynthetic processes of the symbiotic algae, the light exposed surfaces of the incurrent chamber often contain daughter channels in their mucous membrane with symbiotic algae. They can be detected by the darker colouration of these locations. Also the adductor muscle and other organs are frequently covered by a light brownish colouration and house the algae of the species *Symbodinium microadriaticum* in their uppermost tissue layer.

A particularly large specimen of T. crocea (shell length 17 cm).   Photo: R. Wong

5 Clam Diseases

Environmental Damage, Infectious Diseases, and Parasites.

Little is known about the scientific background of infectious and parasitic diseases in clams. This is partly due to the fact that their commercial breeding is a very recent development. The propagation of other species of mussels has a much older tradition, and therefore more research into their pathology has been undertaken.

A thorough knowledge of clam diseases is of utmost importance when releasing artificially reared clams into the wild to enhance populations. Pathogenic germs introduced into the wild population can cause widespread damage, especially in individuals lacking the necessary immunity to combat the disease. Theoretically, propagated populations released into the wild could thus jeopardize the remaining natural stocks.

Even symbiotic algae could cause a threat to the host. They often carry bacteria with corresponding immunity in the clams. Wild populations, however, may lack this immunity and will thus become easy victims of infections caused by these particular bacteria. Therefore research is being undertaken not to ship young clams, but their larvae. The larvae do not yet contain Zooxanthellae. They will then be contaminated with the zooxanthellae symbiotic to the wild population at the particular site of release. It should be possible to cultivate the Zooxanthellae and eliminate the danger outlined above.

A thorough knowledge of clam disease is of utmost importance if clams are to be propagated. As scientists suspect, a high number of casualties among breeding stock attributed to environmental accidents are in reality caused by infectious diseases. Well known pathogens in other mussels and clams like viruses, clamydia, mycoplasmas or fungi, have not yet been described in *Tridacna* species, however, it is hardly imaginable that they should be spared from these hazards. It is suspected that a number of infectious diseases in clams await description.

Commercialization and artificial propagation of clams will certainly boost scientific research.

The pathology of giant clams is a difficult chapter for the aquarist. Little is known about clam disease and therefore diagnosis is difficult. It is possible to monitor the environmental conditions and thus eliminate stress situations, however, to the clams; a lot of symptoms may be hidden in the soft parts under the hard shell and a sudden casualty will leave a multitude of questions. In such a case observation of the other clams in the aquarium is of utmost importance because death does not come without reason. Often only the elimination of possible causes can save the remaining population. To facilitate these procedures I have divided the different expressions of disease in three major groups where the descriptions follow the same scheme. If possible, measures of counteraction are given.

I. Environmental Damage

1. Damage by Heat

Signs: Mantle is not fully extended, stays retracted in severe cases.

Cause: Overheating of water with temperatures in excess of 32–34 degrees Celsius.

According to my experiences individuals of the smallest species, *T. crocea,* and juveniles of other species often show discomfort even at temperatures of 30 degrees Celsius. This holds true under aquarium conditions and it is possible that other environmental conditions can enhance the damage. For instance the gas concentration, according to management of the aquarium water, being different from the state in the sea. At higher temperatures the water contains less dissolved oxygen. An altered salt concentration also influences oxygen concentration. Therefore, with rising temperatures, damage due to lack of oxygen must be considered, enhanced by low atmospheric pressure or high salt concentration. *T.crocea* and juveniles of other species are probably especially vulnerable to the lack of oxygen.

Consequences: Death of clams. The loss will be higher if the period of overheating is prolonged.

Precautions and counteractions: Lowering of temperature, avoid secondary heat damages by low oxygen concentrations (lowering the salt concentration to 1.021 S.G.)

2. Damage Due to Low Temperatures.

Signs: Mantle is partly or completely retracted, slow reaction to touch or light stimulus.

Causes: Water temperature lower than 20 to 22 degrees Celsius.

Consequences: Death of clams. Juveniles are more susceptible to low temperatures than adult individuals.

Precautions and counteractions: Raise temperature to at least 22 degrees Celsius, 25 degrees is even better. As a further precaution, two heater units should be used, each being able to warm the tank individually in case of failure of one unit.

Only transport vessels with heat insulation should be used (styrofoam etc.). Clams are more susceptible to lower than to higher temperatures.

3. Central Bleaching

Signs: Bleaching of the central part of the syphonal lobe, in between the intake and outlet syphon.

Cause and course of event: Loss of symbiotic algae in the area affected. The cause of this, however, is not clear. Braley (1992) suspects a deficiency of a certain spectre in the light source and temperature stress as possible causes. However, I have noticed the phenomenon in animals kept for several years under stable conditions (light source HQI TS 250). Very little is known about the causative conditions. Therefore, I will try to elucidate the condition by some considerations of my own.

Hippopus hippopus, showing heat damage in a clam hatchery caused by a break down in the flow system. Excessive warming by sunshine with a resulting bleaching of the siphonal mantle.

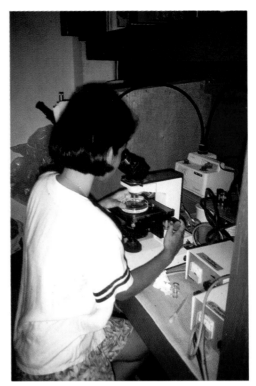

Regular laboratory tests are necessary to monitor the progeny's health

A purely hypothetical explanation could be the clams insufficient ability to produce pigments to shield it from ultra violet light. As a matter of fact, the light hits this central part of the animal vertically, whereas the surrounding parts are more obliquely exposed. The highest degree of damage will therefore be expected in the area of the highest exposure to light. This theory is supported by the fact that the condition develops slowly over a period of weeks or months during constant conditions of illumination.

The animals so damaged seem to be more susceptible to strong illumination than do healthy clams. Animals with pronounced central bleaching will die if exposed to a high-output light source. The clam shown in the picture died after

There is always a cause for the death of a clam.

80

Tridacna squamosa in an aquarium, showing the start of central bleaching.

lenghtening the usual period of illumination by two hours. If the distance to the light source is increased, the condition seems to improve, however, the degenerative process in the tissue will not be stopped, only slowed down. The diminished amount of photosynthesis will further weaken the clam.

If the light source under which the bleaching occurred is not altered, the animals try to avoid exposure by seeking areas of lower intensity of light like hiding the mantle under a soft coral or changing position by abrupt opening and closing of the shells. Whenever I moved the clams back to their original position, the reaction was repeated. If left in their new location, however, the mantle was again opened.

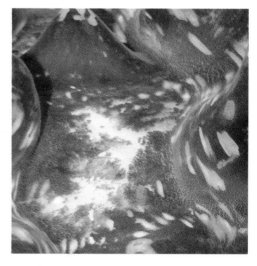

Detail from the clam above.

The condition of bleaching occurs in two distinct forms.

1. Exposure to unphysiologically high amounts of ultraviolet light due to an unsuitable light source with a short distance.
2. A sudden increase in the amount of ultraviolet light, however, which could still be in a physiological range. This could be caused by keeping the animals under low light intensity with a following sudden rise, not allowing the animals to adapt to the new condition. An other condition triggering the disease could be a sudden change in the environment from low to high light intensity, even if both are within the physiological borders for the clam. I suspect that the susceptibility for the condition is enhanced by the long transport in darkness from the tropics to Europe or other countries. After having been kept in total darkness for about 48 hours during transport, the sudden exposure to artificial or daylight could cause a shock. I suspect that this is caused by the UV-rays and not by the other light spectrum which is utilized for photosynthesis.

However, the clams are able to adapt to changing illumination conditions, but the adaption needs time, and must not occur suddenly. Somehow the process is comparable to the tanning of the human skin. A moderate exposure to sunlight and therefore ultraviolet rays triggers the production of a pigment in the human skin which shields genetic information in the cells against damage by further ultraviolet influence. The amount of pigment produced is always in accordance with the amount required for protection without impairing certain biochemical processes, necessary, for instance, to produce vitamin D. A high degree of melanin production, the pigment in question, would block necessary physiological processes but a lowering of melanin production could cause damage to the genetic information stored in the nuclei of cells. In the course of evolution a sensitive regulatory system was developed, allowing for the balance between pigment production and other physiological processes. If the system is stressed by a sudden overexposure to ultraviolet light sunburn will be the result. A long term exposure may lead to a cancer of the skin.

To me, these mechanisms are similar in the *Tridacnidae,* even though the regulatory systems are different. The clams must protect themselves from dangerous levels of ultraviolet light by the production of pigments but on the other hand, the pigment production should not be too high to allow for periods of lower light intensity. Therefore, the amount of ultraviolet light must be measured and pigment production monitored. If tissue damage happens, it will be strongest in the areas of highest ultraviolet exposure, i.e. hitting the syphonal lobe vertically. This is only possible in between the syphonal orifices and this is the location of the central bleaching. From a certain amount of damage onwards, the animals ability to produce pigment will be impaired and the disease will progress. It is likely that due to the tissue damage the channel system of the symbiotic algae is also damaged and they would be lost in the area in question.

In connection with the theory outlined above, a third mechanism could be hypothesized; the pathologically diminished ability to produce protective pigment due to a deficiency in vital substances. As vital substances I mean substances vital for metabolic processes, like vitamins or trace elements like iodid. Certainly the protecting mechanism of clams are dependent on such vital substances but very little is known about it. It is possible that a deficiency of these substances is caused by prolonged periods in captivity, therefore impairing the ability to produce protective pigments. Other changes in the environment like extreme changes in temperature could also influence the animals ability to produce pigment thus causing central bleaching, as supposed by some scientists. So far, the observations are in accordance with such a theory.

Consequences: Unchecked exposure to light will kill the animal.

Precautions and counteractions: Emphasis must be in precautions and should concern all conditions able to reduce the clams wellbeing. Temperature, salt concentration, and pH value must be kept in their respective borders. If the animal is translocated, the necessary changes must be made slowly and carefully looked into. For instance, if an animal is moved from one tank to another, it should happen with as little stress as possible.

A crucial point in the management of central bleaching is the illumination. Light intensity and period of illumination should not differ if an animal is translocated. This won't cause serious problems if the tanks have artificial light, however, problems can arise by taking clams from the hatchery to aquaria. In such instances drastic changes in light intensity are unavoidable. We expose freshly imported clams to a reduced amount of HQI light in the first two weeks. However, the dark period during

transport should be shortened by immediate exposure to light after unpacking, even if that happens during night hours. We recommend a period of 8 hours illumination.

If central bleaching has already taken place, a reduction in ultraviolet light intensity is recommended. This can be accomplished by raising the lamp over the tank to diffuse the rays. An ultraviolet filter can be employed if the spectrum of visibility is not altered. As an emergency measure, the tank can be covered by a glass plate (do not use acrylic!). By doing this, a considerable amount of ultraviolet light will be absorbed. At the same time the daily illumination period should be shortened. It is not only the intensity but also the duration of exposure which causes tissue damage. If the central bleaching cannot be stopped, a translocation of the clam becomes necessary. Also, the regular supply with iodide and other trace elements may be helpful.

4. Local Bleaching

Signs: The localised bleaching in a small area I am describing here, is quite different from central bleaching. Therefore it is mentioned separately. I could find no reference to it in the scientific literature.

Localised bleaching can occur at any site in the syphonal mantle. According to my observations it can evolve rather quickly, over a period of one week. The area is characterised by a complete loss of symbiotic algae but the protective pigment is retained and therefore the spots are not colourless but lose only the brownish tinge. In this respect it is different from the central bleaching where all pigments are lost.

Beginning central bleaching with total loss of symbiotic algae in the tissue affected.

Advanced central bleaching, beginning to become generalized.

Beginning of local bleaching in Tridacna gigas found in shallow water.

Detail photograph of the clam shown above. The tissue shows a total loss of symbionts.

Local bleaching in Tridacna squamosa, caused by translocation from one tank to another with diminished illumination. Bleaching has spread to the fringe of the mantle. The retained protective colouration can be clearly seen.

Cause and course of the condition:
I was able to observe this condition several times after a reduction of light intensity without changes in light quality, duration of illumination, or other environmental conditions. A change of location in an aquarium alone was enough, in some instances, to trigger the condition. In another case, the elimination of certain spectrums of light caused the disease, without any change to ultraviolet light intensity.

It seems that the condition is caused by a reduction in the number of symbiotic algae but not a primary damage to the mantle tissue. Therefore we are dealing here with damage to the symbiotic algae but not to the animal. I was able to observe the same phenomena in *T. gigas*.

After intensifying and prolongation of the illumination, regeneration of the brownish colouration occurs, beginning at the edge.

Close-up from the same area.

The animals were kept in a clam hatchery in the Philippines. It was the rainy season (June to October), and the animals had to cope with a reduced light intensity due to heavy clouds after the bright sunshine of summer (March to May). Though the animals are kept under seminatural conditions, the environment is partly dictated by human interference. Normally they were kept at a depth of about three meters which must be considered shallow for *T. gigas*. Under natural conditions they are found in much deeper water. With the onset of the rainy season with its diminished light intensity, the clams may be forced to survive on the border of their adaptability. However, local bleaching in the hatcheries was only observed in *T. gigas*, but not in other species like *T. crocea, T. derasa, T. maxima, H. hippopus* and others, their natural habitat being in higher areas of the reef than that for *T. gigas*. Specimens of the latter, found in 18 m depths, never showed signs of local bleaching.

Consequences: In serious cases of local bleaching the condition can spread over the whole area of the syphonal mantle and a generalized bleaching will result. In other words, a loss of the entire population of the symbiotic algae. It is interesting to note that the iridescent protective pigmentation is not affected, leaving a distinctive pattern on the mantle. The designation "local bleaching", however, is still justified because a generalization of the process is a rare complication. In the case of the classical local bleaching, the animal can recover after several weeks to months but the generalized form will inevitably lead to the animals death. The photographs show the progress of the disease. The recovery of the symbiotic algae does not occur in a random distribution but always starts from the fringe of the mantle.

Prevention and counteractions: A reduction in light intensity should be avoided, and if inevitable, should be carried out in a protracted way. Changing from daylight fluorescent lamps to blue types will always cause a reduction in light intensity because of the differing spectra. Such a change in illumination must be carried out over a period of several weeks changing from only one type of tube at a time.

A more specific remedy cannot be given and the recovery of the symbiotic algae population is mainly a question of time.

5. Generalized Bleaching

Signs: uniform clearing-up of the syphonal mantle. The rather dark brownish colour brightens. Microscopic slides show a reduction of the symbiotic algae.

Causes: Lack of nitrogen in the tank water. The number of symbiotic algae per cubic mm in Clams is about 10 times higher than in corals, due to their specialised circulatory system (Carmen Belda, pers. comm.). The result is a much higher nutritional requirement, and therefore a higher amount of organic nitrogen will be taken from the environment in comparison to the same surface unit in corals. In aquaria with a high density of these molluscs, nitrate as well as phosphate levels can become low. I experienced a dramatic lowering of nitrate levels in an aquarium with a dense clam population over the time, reaching non-measurable values. Corals, also present in the tank (species of *Acropora, Pocillopora* etc.) showed a remarkable growth rate during the time due to the low contents of organic nitrogen. However, the growth rate of the clams came to a standstill, clearly shown at the fringe of the shells and a few months later, a significant general clearing-up of the syphonal lobe was observed followed by the unexplained death of a number of individuals, always the ones with the highest degree of general bleaching. Signs of other forms of bleaching already described and attributable to conditions of light were not observed.

Consequences: If the nitrogen deficiency is not remedied, the photo-synthesis will be drastically reduced and the animals will eventually die.

Precautions and counteraction: As a preventive measure, the nitrate contents of the tank water must be regularly monitored. Nitrate is only one component of the many organic nitrogen compounds but it can most easily be measured under aquarium conditions. In the case of a low nitrate concentration either the number of clams must be reduced or nutrients must be added to the tank water. For aquarists who do not want to keep either more fish or artificially add more nitrate, it is good advise to lower the number of clams kept, especially large specimens.

Another possibility is a light fertilization with a 0.1 % solution of sodium nitrate (analytical quality; Merck). One gram of sodium nitrate to 1000ml of distilled water will give you the stock solution. Not more than 10 ml per 100 l tank water should be applied and the nitrate contents measured. Highly accurate measurements must be employed. Ammonium nitrate could also be used but bears a higher risk of poisoning when overdosed.

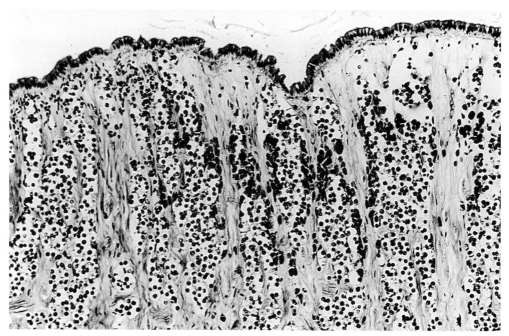

Microscopic slide of the siphonal mantle with a normal density of symbiotic algae.

Photo: Dr. John Norton

Generalized bleaching caused by degeneration of symbiotic algae.

88

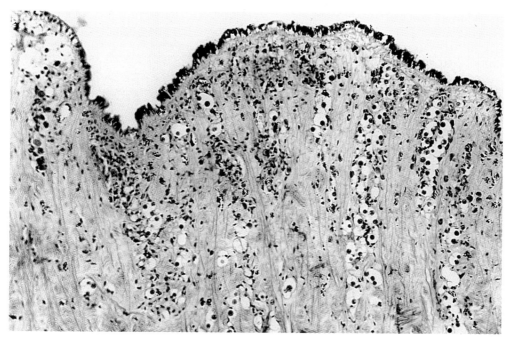

Microscopic slide showing mantle tissue of a clam with bleaching. The number of symbiotic algae is pathologically low.   Photo: Dr. John Norton

◄

Tridacna crocea with loss of symbiotic algae, caused by a deficiency in nitrogen compounds.

6. Damage Caused by the Lack of Light.

Signs: Uniform bleaching of the whole mantle, in the case of shadowed parts, partial bleaching.

Cause and course of event: Local bleaching as described above is caused by acute lack of light. Longer periods of low illumination, however, do not lead to circumscribed areas of bleaching but to a retardation of physiological processes. The clam tries to adapt to the new situation by "breeding" more symbiotic algae. At the same time the life expectancy of the individual algae is prolonged. As a consequence the algal density will be higher, resulting in a deep brown coloration of the mantle. The amount of protective pigmentation compared to ultraviolet light is also diminished as a consequence of the lowered overall light intensity. The iridescent pigments would prevent light from reaching the symbiotic algae. To catch as much light as possible, the clam will extend its mantle as far as possible, a fact often misinterpreted by inexperienced aquarists as a sign of well-being. When the animal dies eventually, the astonishment is great, and a predator, for example a polychaete, is suspected as the culprit.

If the light intensity drops further, the symbiotic algae will die. This will happen first in the region of the lowest light reception, i.e. in the parts of the mantle which do not receive the light vertically. This is a disadvantage of the artificial illumination, always coming from the same direction, in contrast to the ever changing natural illumination by the sun in the coral reef. Because of the folded architecture of the mantle, parts of it may be in the shadow, but only for a while. With the changing position of the sun, conditions will be altered. Under aquarium conditions, however, parts of the animal may be permanently shadowed with consequential bleaching of this particular area, and loss of symbiotic algae. I experienced this phenomenon in an adult *T. squamosa* (shell size 32 cm) on both sides of the inlet syphon and behind the outlet syphon. In this species the outlet syphon slopes anteriorly, and therefore shades most of the remaining tissue.

The same situation can occur when the clams stand very close together, probably caused by a high growth rate, where the mantles shade each other. In such cases a partial or complete loss of symbiotic algae can follow.

In all cases of partial shadowing, however, the loss of algae only occurs in the parts with diminished light intensity and does not affect other areas of the clam and the parts are not as sharply outlined as in the cases of local bleaching, already described. In contrast to local bleaching, the condition proceeds slowly. If the illumination is of such low intensity that the animal cannot compensate for the deficiency, the whole of the mantle will be discoloured. At first the protective coloration will fade, followed by an intensified brown colouration caused by a higher number of symbiotic algae as a compensating factor but eventually the brownish colouration will also fade due to the death of the algae. This mechanism will, however, not lead to the immediate demise of the animal, at least not until the stored nutrients are utilised. The animals are able to withstand certain periods of lower light intensity as this occurs in the natural habitat during the rainy season. When the rainy season is over, the gradual rise of the light intensity will also crease the number of symbiotic algae.

Consequences. If the light intensity over the aquarium is not be intensified, the discolouration of the mantle will continue, and the animal will eventually die.

Precautions and counteractions: If a partial bleaching of the protective colouration without a distinct border is observed, a chronic lack of light must be considered. Without moving the animal, a stronger light source should be employed, sometimes it is enough to lessen the distance of the lamp to the water surface or cleaning the cover glass from crystallized salt. The animal, however, should be moved if it located at a permanently shadowed location.

The same measures should be taken in case of a general discolouration of the mantle. In such cases the animal is lacking its protection from ultraviolet light and the increas of the light intensity should be carried out slowly over a period of time.

7. Heavy Metal Poisoning.

Signs. Complete bleaching of the clam or sudden death.

Cause and course of event: As all molluscs, clams are very sensitive to heavy metals. Heavy metals in diminutive amounts are a natural ingredient of seawater and necessary for certain biochemical processes. In higher concentrations, however, they can prove fatal. Two different reactions in clams to heavy metals can be observed. They are not specific but depend on concentration levels. Very little is known about the physiological background of heavy metal action in clams. Future scientific work may reveal specific reactions to different components.

First sign of heavy metal poisoning is the development of a lighter colouration with the loss of the dark brown colouration, obviously a reduction in the number of symbiotic algae. In such a case the heavy metal concentration is such that it kills the algae but not the clam.

The dead algae will be phagocytized by the clam's defensive system and consequently will be incorporated into its tissue. Therefore, any general discolouration should arouse the suspicion of heavy metal poisoning.

Consequences: All molluscs are sensitive to heavy metal poisoning and the prognosis is poor. If signs of heavy metal poisoning are recognized, it is in most instances too late. Only once I succeeded in rescuing one of three *T. crocea* after severe poisoning, by transferring it to a new environment where it flourished thereafter for many years.

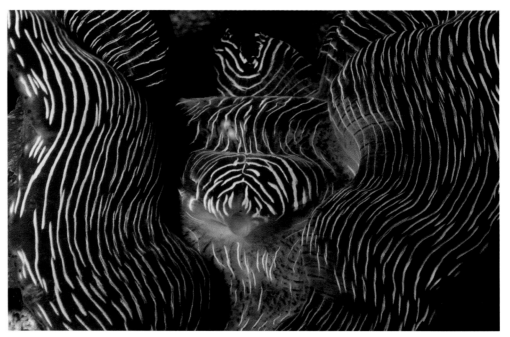

Tridacna derasa with moderate light deficiency damage. In the shadowed area under the excurrent siphon and anterior mantle loss of symbiotic algae. Higher density of symbiotic algae in the light exposed parts of the mantle.   *Photo: Svein A. Fosså*

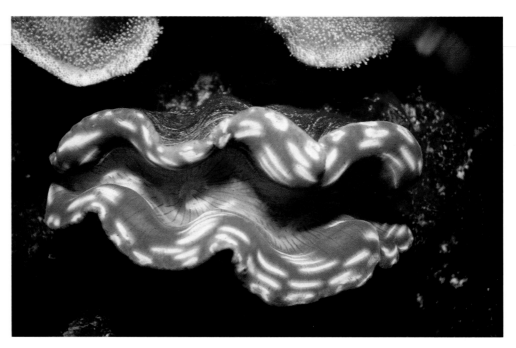

Tridacna derasa with bleaching due to light deficiency in the shadowed area of the middle part of the mantle.

Bleaching of shadowed areas in Tridacna maxima. Darker colouration of light exposed parts due to a compensatory higher density of symbiotic algae.   Photo: Svein A. Fosså

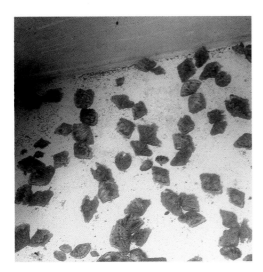

Poisoning can cause the immediate loss of many clams. The picture shows dead clams after Endrine poisoning in a hatchery on January 12th, 1992.
      Photo: Silliman University, Marine Laboratory

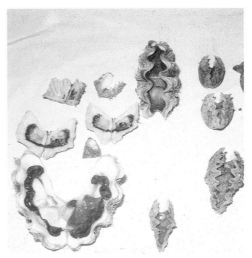

Larger specimen poisoned by Endrin. Poisoning was caused by polluted sea water.

Prevention, counteractions: Several conditions can cause a high heavy metal concentration in the tank. You have to think of pumps with exposed metal parts (except stainless stell), unsuitably framed aquaria and residues from medicaments to treat fish diseases like copper sulfate. Heavy metal contaminations in the tap water also must be considered, even the water from rozen food should not be brought into the tanks. Other possibilities should be thought of and the aquarist should thoroughly check the whole system. In most cases, however, the effect is so dramatic, that help comes mostly too late. Only immediate transfer to a heavy metal free tank can sometimes save some of the poisoned animals.

8. Damage Due to Hypersalinity

Signs: Even during illumination, the shells are not completely opened and the mantle stays mostly retracted.

Cause: Salinity too high.

Consequences: Further raise of salinity will cause death.

Prevention and counteractions: Thinning of salinity by adding freshwater. Regular salinity checks and topping-up of evaporated water. Salinity should be measured with a refractometer or hydrometer

9. Damage Due to Hyposalinity.

Signs: In case of low salinity, the shell will stay closed and the mantle retracted.

Cause: salt concentration too low.

Consequences: Inflammation of gills. Longer duration and further lowering of salinity will cause death.

Precaution, counteraction: Slowly raise salinity. Regularly check salinity with a refractometer or hydrometer.

10 Gasbubble Disease.

Signs: Gasbubbles in tissue of syphonal lobe. Smaller animals with lighter shells may even float. There is no inflammatory reaction around the gas bubble.

Cause: Gas contents of tissue. Scientific investigations showed that saturation of the water with atmospheric gases is responsible of the condition. The gases are then concentrated in the syphonal lobe forming bubbles (Bisker &. Castagna, 1985, Malouf et al., 1972) Keeping the clams in the large tanks of the hatcheries may cause the condition as that is where it is mostly observed.

The temperature of the water could also play a role. High temperature means higher gaseous pressure which then could cause bubbles in the delicate tissue of the animals if abruptly transferred from colder to warmer water. It is also possible that pressurised water in the pumps could rise gaseous concentrations, leading to the condition if animals are abruptly transferred to such an environment. As an example a pump is mentioned with a leak in the tube sucking air into the system, similar to a jet pump (Braley 1992).

Consequences: Serious cases cause death.

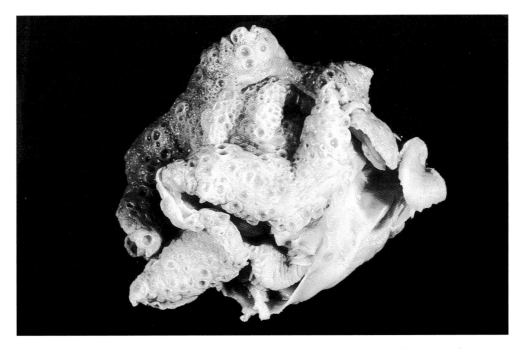

Fully developed gas bubble disease (Tridacna gigas, 8 cm length).   Photo: Dr. John Norton

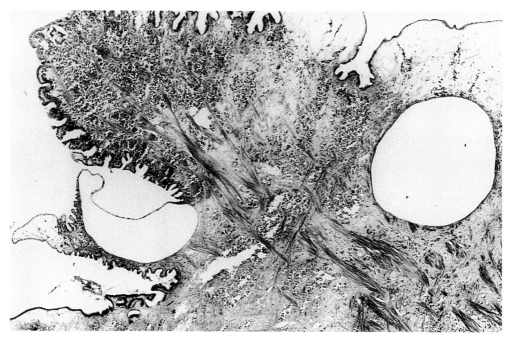

Microscopic slide, showing tissue of a diseased Tridacna gigas (3 cm). The gaseous bubble to the right of the picture is clearly seen. Photo: Dr. John Norton

Precautions, counteractions: Every change in environment must be cautiously carried out. Clams are able to adjust to differing gaseous pressures in the environment, but this is a physical process and takes time. This is a process similar to the Caisson disease in SCUBA divers. If decompression times are not obeyed, gaseous nitrogen will form in tissues and circulatory system with dramatic effects. We allow our animals plenty of time to adapt to a new environment. If the condition is obvious, the prognosis is poor. If, after translocating an animal, bubbles start to show in the syphonal lobe, it should be carefully relocated.

II. Infectious Diseases

1. Bacterial Damage to Tridacna Larvae and Juvenile Clams

Signs: Destruction of the mantle, sudden death (48 h).

Cause and course of condition:
Bacteria of the *Vibrio, Xeromonas,* and *Plesoimonas* group. They are gram negative. These bacteria flourish in hatcheries if water hygiene is lacking or in case of overfeeding. Especially dead clams are an excellent breeding ground for them and must be removed immediately, otherwise healthy animals may be infected. Cultures of unicellular plants, utilised as food for *Tridacna* larvae, also host several species of bacteria which are dangerous to molluscs.

Consequences: Usually the animals die in 24 to 48 h.

Precaution and counteraction:
Nothing can be done if animals are infected because they die before an antibiotic will take effect. As a precaution, thorough water hygiene with removal of sediment and food residual is recommended. Water should be changed in regular intervals. The general environmental factors like salinity, pH-value, temperature, and illumination must all be in their respective tolerances. Suitable substrate must also be provided on which the animals can insert their byssus apparatus.

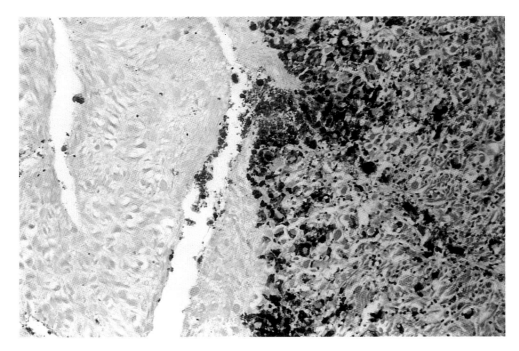

Microscopy of a bacterial infection with subsequent destruction of tissue in the adductor muscle of Hippopus hippopus (14 cm).   *Photo: Dr. John Norton*

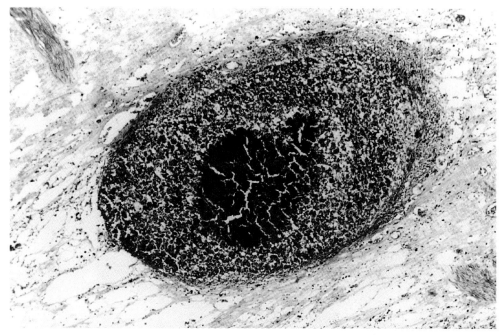

An abscess in the siphonal lobe of a dying Hippopus hippopus (30 cm), caused by bacteria.
Microscopic slide.   Photo: Dr. John Norton

2. Bacterial Iinfections of Juvenile and Adult Clams

Signs: Shells not fully opened, mantle partly retracted. Sometimes yellowish abscesses are visible on the syphonal lobe with destruction of tissue.

Cause and course of condition:

Infections with Vibrio species, widely distributed in the marine environment, are the likely causes. Microscopically, the necrotic tissue shows colonies of gram negative bacteria. They are also found in the lymphatic fluid which is normally sterile.

Mainly the condition can be traced to stress factors in the history of the individual clams, e.g. temperatures which are too high or too low, stress during transport or sudden changes in other environmental factors. All these conditions weaken the immune system of the animals making them susceptible to infections.

Consequences: Death of the animal.

Precautions and counteraction: A treatment with drugs is not possible at the moment. As a precaution any stress situations should be avoided. During transport, temperatures should stay constant and every change in environment should be carefully carried out.

3. Infections by Rickettsia

Signs: None

Cause and course of condition:
Preconditions are not clear. However, Rickettsia flourishes better under aquarium conditions than in the coral reef and is therefore a higher risk to clams in the aquarium than in the natural habitat.

The disease is caused by Rickettsia. Systematically they stand between bacteria and viruses. Microscopically they show intracellular basophilic inclusion bodies. They are named after the American pathologist Howard T. Ricketts. Mainly they colonise the outer layers of the gills, forming microscopic cysts. They are also found in the lateral mantle, in the vicinity of the tissue layer attached to the shell. Chlamydia is also often present in the outer tissue layers without causing disease. They can be distinguished from Rickettsia only with the help of transmission electron microscopy.

Consequences: In case of a heavy infection, the animal will die.

Precaution and counteraction. Nothing is known to prevent the disease. To prevent the spread of infectious disease, animals appearing affected should be isolated and dead clams must be removed immediately. Permanent use of ultraviolet light sterilisation is recommended.

4. Infection by Perkinsus

Signs: None

Cause and course of infection:
Infections by species of Perkinsus. They are normally found in the digestive tract. No typical symptoms for the infection are known but it seems likely that they are responsible for great losses in hatcheries and affecting not only clams but other commercially bred mussels as well (Luckner 1983).

Consequences: These are not clear. Perkinsus has been found in the reef on dead as well as on life mussels.

Precaution and counteraction: not known.

5. Infections by Protozoa.

Signs: None in the living animal. The dead clam shows whitish spots on the kidneys.

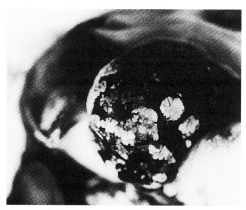

Whitish discolouration of the kidneys in protozoal infection, caused by a Marteilia related species. (Tridacna maxima, 14 cm).
Photo Dr. John Norton

Cause and course of disease: The disease has only been described from populations of *T. maxima* from the Fiji Islands. The causative agents are species of *Marteilia* or *Paramarteilia*. In mussel species other than giant clams infections have caused great losses.

Consequences: Destruction of the kidney tissue with following death of the animal.

Precaution and counteraction: Not known.

6. Whitespot Disease *(Knop/Norton)*

Signs: White round spots on the syphonal lobe. In cases of slight infection, the shell is fully opened, however, if the disease progresses, signs of discomfort become obvious. The

Tridacna crocea with first signs of "WSD" infection.

clam is only partly opened and most of the mantle is retracted.

Cause and course of disease: This disease has only recently been

Start of "WSD" infection on the syphonal mantle of Tridacna maxima.

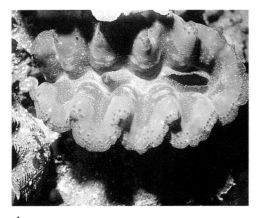

▲
Tridacna crocea, heavily infested by the highly contagious "WSD" disease.

▶

The whole mantle of a T. squamosa. At this stage of "WSD", the clams do not open completely, hinting at a reduced general condition.

▼

Aquarium specimen of Tridacna squamosa (32 cm), with "WSD". The differently sized cysts represent parasites in different developmental states.

described in the family *Tridacnidae* and has very distinct signs. The first description was made by myself and is based on infected clams owned by me and a local dealer (Knop 1994 B). The existing photos showing the symptoms were taken in my aquaria. All species kept in the tank were affected, namely *T. crocea, T. maxima,* and *T. squamosa.* It is likely that other species could be infected as well. The species infected in the dealer's tank mentioned above was *T. crocea.* It is possible that the simultaneous occurrence in my aquaria and these of the dealer are interrelated.

The disease seems to be highly infectious as more than 40 clams in my tanks fell victim to the infection. Some clams seem to have strongly fought the disease while others showed no signs till the end. A reduced immune response is likely to have contributed to the epidemic-like spread. The animals had just survived the extremely hot summer of 1994 and their symbiotic algae had been reduced due to a lack of organic nitrogen which is indispensable for the *Tridacnidae.* All this may have contributed to a reduced resistance.

The small, whitish cysts, characteristic for the disease, are of different sizes, probably a hint for different stages in the development of the causative agents. At the beginning only few cysts are visible, causing hardly any discomfort in the clam. If slight pressure is put to the tissue surrounding the cyst, a small whitish spherical body is expelled. Larger cysts may expel this material already after a sudden contraction of the syphonal mantle. In cases of heavy involvement, cysts are not only found on the surface but on the underside of the overhanging part of the mantle and the site of the byssus aperture. The lateral mantle and the tissue section covering the kidneys is also often affected. It is if distinctive appearance and cannot be mistaken as Marteilia or Paramarteilia infection.

The spread of the disease in a single tank lets one think of the possibility of a planktonic stage in the development of the supposed parasite but I was never able to identify such a free swimming phase. Such a reproductive cycle is well known in many parasites pathogenic to fish. I never succeeded to check the spread of the disease by water sterilisation.

The Australian Tridacna specialist and veterinary pathologist, Dr. John Norton, carried out electron microscopic studies in diseased clams and found species of Protozoa as the causative agents. This finding was confirmed by the American specialist in Protozoa, Prof. Frank O. Perkins, the discoverer of Protozoa of the Perkinsus-type. He recognizes similarities to infections with *Perkinsus atlanticus* and *P. olensi* (Perkins, F. O., 1985), therefore the White Spot Disease seems to be caused by a species of Perkinsus.

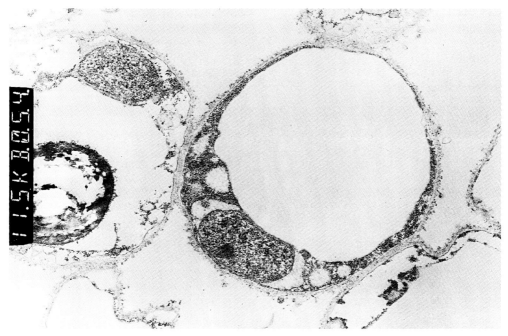

Transmission electron microscopy of the "WSD" causative agent. Magnification 23 000.

Photo: Dr. John Norton

Consequences. None of the infected animals survived. The period spanning the time from the first recognition of the disease to the demise of the animals differed from a few weeks to three months. Large animals proved to be hardier than small ones, irrespective of age. For instance, juveniles as well as adults of *T. crocea* are easy prey to the disease whereas larger species like *T. squamosa* (overall length 32 cm) seemed to be much more resistent.

Precaution and counteractions: Not known. Spectra and intensity of light do not seem to have any influence on the infection. At the time of infection, the water temperature was high with about 30 degrees Celsius and it is possible that a lower temperature of 25 degrees Celsius may enhance the clams resistance. D. Koutroumbilas reported about totally cured clams after antibiotic treatment with Chloramphenicol (1996, pers. comm.).

III. Damage from Macroparasites and Fish

1. Snails

a. *Muricidae*

Signs: These are many because a multitude of snails can be responsible for the damage, each species having its specific feeding behaviour. Soft parts of the clam can be attacked but also trepanation of the shell occurs.

Cause and course of events: The group of murexine snails being noxious to clams is large. It comprised species of the family *Chicoreus (C. bruneus, C. microphyllum, C. ramosus), Cronia (C, fiscella, C. margariticola, C.*

ochrostoma), Morula granulata, and *Thais aculeata.* The mode of feeding is reflected in the number of species, as already mentioned. *Chicoreus brunneus* and *Cronia fiscella,* for example, are drilling wholes into the shells utilizing calcium dissolving acids to gain access to the soft parts. Others like *Chicoreus ramosus* use the Byssus aperture or other natural openings of the shell. The Byssus opening is especially vulnerable if not closed with stones like over sandy bottom or broken coral.

Consequences: Though an attack by muricidae snails can be lethal for clams, mass killing is rare. In the aquarists tank mostly only one individual will be involved and the loss will not be serious.

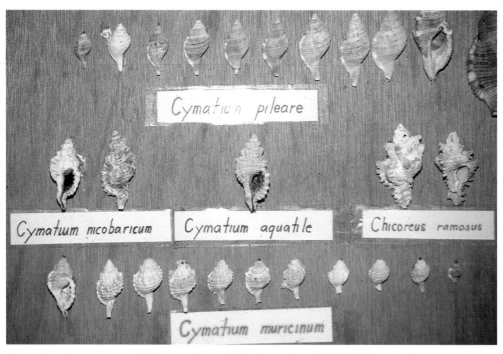

Several snail species, parasitizing on clams. (Collection of the UP MSI, Philippines).

Chicoreus ramosus enters the clam mainly through the byssal orifice.

Precaution and counteraction:
Regular control of the clam population and removal of suspicious snails is the proper precaution. Every clam newly introduced into an already existing population should be thoroughly checked and even small snails should be removed.

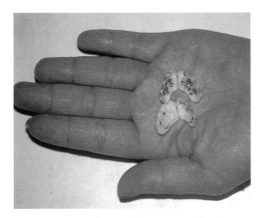

Many parasites deposit their eggs on the inner surface of dead mussel shells.

b. *Pyramidellidae*

Signs: Mantle not fully extended or partly retracted and the shell not completely opened.

Cause and course of event: Attack by pyramidelline snails. These small parasitic snails are widely distributed, covering the distribution of the clams. They are numerous in Australia, Guam, Indonesia, Palau, the Philippines, Papua-New Guinea, and the Solomon Islands. Twice I experienced contamination of aquaria with these parasites, in each case the snails could be traced back to mussel hatcheries.

All species of the family comprise small snails with an overall length of a few mm only. In most cases *Turbonilla* species are responsible for the damage but *Tathrella iredalei* was also observed parasitizing clams (Braley 1992). Mostly the parasites are found on the outer fringe of the shell, hiding under the overhanging mantle. They feed at night on the hosts lymphatic fluid. If only one or two parasites are present, the damage would certainly be in tolerable borders but if the snails occur in large numbers the clams health would be seriously impaired. In the field I found hundreds of these snails parasitizing on large clams.

The snails find a ready environment in closed systems as used in clam hatcheries or aquaria, the reason for this being the large number of clams and the absence of predators like small fishes of the family Labridae. The reproductive cycle only takes two weeks though the metamorphosis from the egg to the the final snail is a complicated process incorporating planktonic larvae. Sexual maturity is reached in 40 to 50 days and the life span is about four to five months. Obviously the snails are able to

This picture shows Pyramidelline snails "working" on the mantle of a juvenile Tridacna gigas. Clearly the trunk can be seen sucking lymphatic fluid from the mantle. Length of snails about 6 mm. Photo with kind permission of the James Cook University, North Queensland, Australia.

▼

The comparison with the lineal scale demonstrates the small size of the snails.

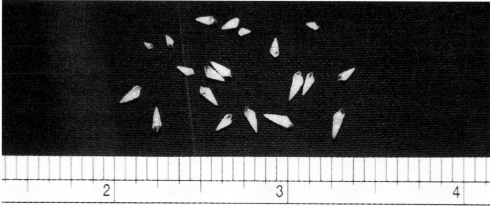

SCIENTIFIC SUPPLY (

reproduce in captivity if certain requirements are met.

Consequences: Spread of these snails in the aquaria can pose a serious hazard to the clams by reducing their defense mechanism. Without countermeasures the clams will eventually die.

Precautions and counteractions: It is of great importance, in the case of parasitic snails, to manually remove the parasites. Preferably this is done at night as this is the time the snails are most active and are leaving their hiding places. A thorough search should not only include the shells mantle area but also the rest of the shell and the substratum. Another preventive measure is the introduction of a predator feeding on the snails like a small individual of one of the wrasses. In my experience Coris gaimard is perfect for this purpose. As long as this species still shows the juvenile colouration which is a bright orange, they will not pose a thread to the clams. When the adult blueish colour starts to appear, however, they are better removed from the tank. This can be quite a task because of their ability to dig themselves into the sandy bottom and it is equally difficult to catch them in traps.

Another measure checking a pyramidelline threat is to interrupt the reproductive cycle of the snails, preferably by removing the planktonic larvae with the help of a powerful skimmer in connection with ultra violet light disinfection. A large colony of clams in an aquarium could be thus freed from a heavy infestation with snails.

c. *Other Species of Snails*

This is a large group of snails comprising species of the *Costellaridae (Vexillum plicarium, V. cruentatum), Ranellidae (Cymatium aquatile, C. muricinum, etc.) Fasciolaridae (Pleuroploca trapezium)* and *Buccinidae.*

The layman is not advised to get too much involved with these species in as much as other snails are also suspected to be of parasitic nature and the list will certainly be extended in the future. It is good advise to remove all snails from the tank, even if now and then a harmless algae feeding species will be included. For the sake of other sessile inhabitants of the tank, snails should only be tolerated if their harmless herbivorous feeding behaviour is known. Carnivorous snails must not be tolerated in the coral reef aquarium. However, some snails are omnivorous,

A further species of snails parasitizing on tridacnids was discovered in the aquarium. On the underside of dead specimen of Tridacna crocea, having made a quite healthy impression the previous day, more than 50 snails per individual were found. The snails were about 8 mm long.

being herbi- and carnivorous and consequently it is best to avoid all snails in the tank. As already mentioned, *Coris gaimard* or similar species can be used as snail eaters, but as they grow they can also be a threat to the clams.

Many Cymatium species parasitize on clams.

The feeding behaviour differs in different species of predatory snails. In the picture, a juvenile clam is being devoured by a gastropod.   Photo: Marine Lab, Silliman University

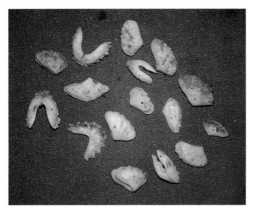

In many dead juvenile clams from the hatchery, bore holes hint at predatory snails.

The house of the predatory snail Cymatium aquatile.

Cymatium muricinum, another snail parasitizing clams.

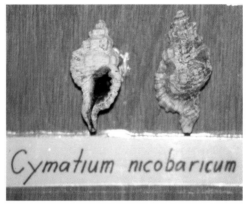

Two species of tritons of the Ranellidae family.

These harmless, "herbivorous" species are employed in clam hatcheries to reduce algal growth.

A parasitizing snail was surgically removed from this Hippopus porcellanus, having worked itself far into the clam.

2. Flatworms

Signs: Being endoparasites, they can only be observed by dissecting the deceased animal.

Oysters and mussel are parasitized by several species of Planaria (Littlewood and Marsbe 1990), but in the *Tridacnidae* only one species of the family *Stylochidae* has been described and only in *T. gigas*. The affected clams were kept in floating cages in the open sea to avoid infestation with snails but of course an attack by free swimming flatworms could not be avoided. The resuming loss was considerable.

The flatworms, being very small and of brownish colouration, are effectfully camouflaged on the mantle surface and not easy to detect. In case of the death of the host, however, they can be found inside the clam. Their size range is from six millimeters to six centimeters. Nothing is known about the geographical distribution of the parasites and they probably also do occur in other regions than the coral reefs. Therefore, they may be encountered in an aquarium populated with non tropical organisms.

It is supposed that the flatworms enter the clams through the byssus orifice and after the *Tridacna sp.* has died, large masses of eggs are released, carrying the parasits to other clams.

Consequences: Death of clam.

Precaution and counteraction: None. In case of a clam's death, only dissection can prove the presence of the flatworms. Of course the tank should be thoroughly cleaned.

3. Boring Sponges

Signs: Small holes of 0.5 to 1.5 mm diameter in the clam's shell, containing the boring sponge. The holes are distributed in groups. If the mantle of an affected dead clam is removed, a network of delicate channels will be revealed, containing also sponge tissue. Near the surface, the bright colouration of the boring sponge, mostly orange, brown, green or yellow, can be recognized (Thomas 1979). In cases of heavy infestation, the shell's surface bulging outwards.

Cause: Infestation by boring sponge. However, certain algae and worms cause similar pictures.

Consequences: Without treatment, the animal will be considerably weakened, making it susceptible to other diseases. It is not clear wether the sponge itself is able to cause death or if it only provides a port of entry for other parasites.

Precaution and counteraction: A clam showing signs of boring sponge infestation must be removed from the tank. The shell should be brushed with a one per cent formalin solution and kept outside the water for about one hour for optimal effect of the treatment. The clam will not be harmed by being left dry. In the natural habitat, the same often happens due to the action of the tides, leaving the clam in the hot tropical sun. After the treatment, the formalin is removed with freshwater and the clam returned to their habitat. It should be mentioned that this sort of treatment is used in the professional clam hatcheries and under home aquarium conditions the formalin should be removed with great care. The animal should be then kept in a separate tank for some time to get rid of the last traces of formalin.

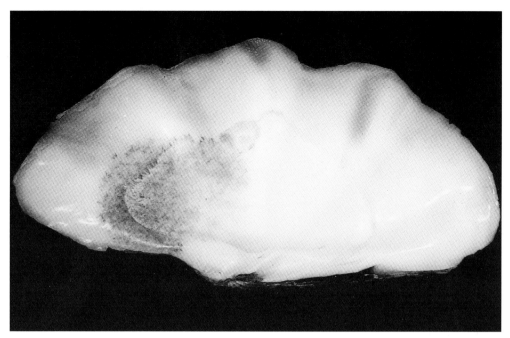

This shell valve of a T.gigas (30 cm) shows a massive infection with a boring sponge (Cliona sp.).
Photo: Dr. John Norton

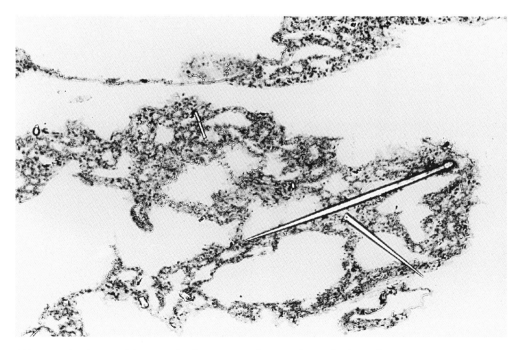

Microscopic picture of the same boring sponge, showing the sceletal needles.
Photo: Dr. John Norton

111

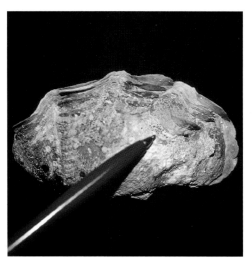

The sponge bores holes into the shell's surface.

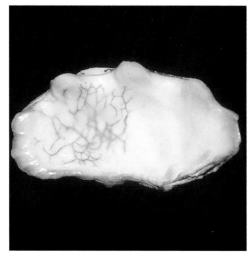

Meshwork of ducts produced by the boring sponge in the same individual (Tridacna maxima of 10 cm), seen from the inside.

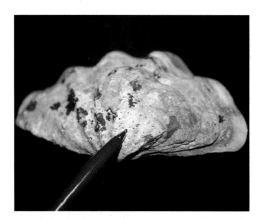

Bore holes at the lower and therefore older part of the clam indicate an early infestation (Tridacna derasa, 13 cm).

The inner surface of the same individual. Apart from ducts caused by the boring sponge, a spread of boring algae is also noted.

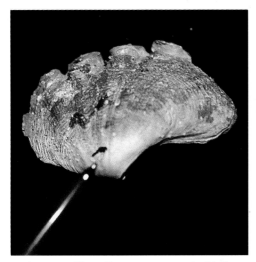

Holes caused by other parasites are the likely sites of entrance for boring algae.

A massive attack of boring parasites killed this Tridacna derasa.

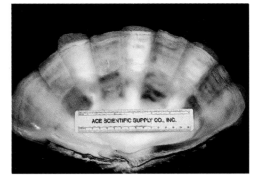

Inner surface of an aquarium raised Tridacna squamosa (32 cm) with a heavy boring algae infestation.

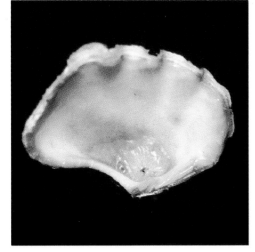

The inner surface of the same clam shown above (Tridacna crocea, 7.5 cm). Heavy infestation with boring algae in the vicinity of the artificial gap. Green colouration at the fringe of the shell are caused by algae entering the clam through natural openings.

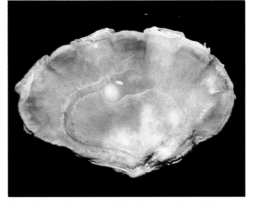

Final state of an attack by boring algae. The whole of the inner surface shows green colouration.

4. Boring Algae

Signs: Greenish colouration of the inner surface of the shell with irregular outlines. In cases of heavy infestations, one or both shells are completely involved.

Cause: Though the phenomenon is well known with clam specialists, the causes are not clear. Even well known veterinary pathologists and biologists in Australia and Asia could not provide an answer to the question. I will give my own impression of the disease until further knowledge will elucidate the condition.

I suspect shell penetrating algae, entering the clams through their natural openings but possibly also penetrate the shells. However, the holes are so small that they cannot be seen with the unaided eye. In numerous clams I found a greenish discoloration of the inner fringe of the shell with a sharp border to the pallial line. I suspect that the algae enter the shell at the before mentioned site, possibly aided by only partly opened shells. If the mantle is not fully extended during daytime, for what reason ever, more light will be shed on the shell's fringe and therefore enhance algal growth. If they have reached the inside of the shell, they will further penetrate the calcium. I have measured an algal growth to a depth of 0,5 mm. In the case of pre-existing holes, they also could provide sites of entry.

I even recognized cases where the whole thickness of the shell showed the greenish discolouration, especially in juvenile individuals with thin shells. The lower parts of the shells were mostly involved whereas the higher and younger portions of the shell were not affected.

Consequences: I do not know if this algae are truly parasitic because a negative effect on the animals health has not been verified. In *T. derasa* I once found an individual whose lateral mantle was nearly completely missing and with heavy growth of the algae in both shell valves, but it was not clear wether the event was caused by the algae or other noxious organism, for example polychaetes. It could be possible that the algae themselves are harmless but could facilitate attacks by other parasites.

Precautions and counteractions: No natural enemy exists for this algae under aquarium conditions and the closed system of an aquarium provides excellent conditions for their growth. It was not possible to control growth by a strong ultraviolet sterilisation. Nitrate and phosphate concentration seem to have an effect on the growth rate, but they were also found in tanks with a low nitrate concentration causing bleaching of the clams. It seems advisable to isolate affected animals.

Do not purchase animals with signs of these algae.

5. Crabs

Signs: Tissue damage in the vicinity of the mantle or the byssal orifice. Large chunks of tissue can be missing.

Cause and course of event: Parasitizing crabs. Several species of crabs, mainly members of the *Xanthidae* and *Portunidae* are responsible but hermit crabs *(Dardanus species)* can also be harmful and the damage they can cause in hatcheries is considerable, especially in juvenile clams. In the aquarium the parasitizing species of crabs are normally rather small but larger species do occur belonging mainly to the *Dromiidae.* If still small they survive on remnants of food particles but adults will attack the clams for want of other food stuff even if clams are not a constituent of their natural diet. Therefore, all the crabs mentioned above must be treated with suspicion.

Consequences: Larger crabs will force the shells of clams open, especially juveniles in hatcheries are at danger. This hazard will rarely be encountered in the aquarist's tank, however, if damage occurs with a suspicion of crabs being the cause and in the absence of fishes they should be thoroughly searched for. If the damage to the clam is only small and the crab can be found and removed, the animal will have a good chance to recover. Otherwise, the prognosis is rather bad.

Precaution and counteraction: It is most important to remove all crabs being suspicious of a parasitic behaviour. Symbiotic crabs living on stony corals like Acropora are harmless and are mostly confined to the coral as habitat. In addition, they are very small and thus do not pose a thread to the clams. Other crabs should be always considered dangerous but the task of removing them from the tank can be a difficult one. Some crabs (*Mithrax* spp.) can sometimes be trapped with nylon stockings after being baited with mussel meat. The crabs will become entangled in the nylon and thus get caught when attempting to get hold of the meat at night. This trick, however, does not always work and other species of crabs hardly ever fall to it. In such cases crab traps should be employed. They are available at pet shops.

Damaged clams should always be removed from the tank. This, however, means a change in environment which can cause a further hazard.

A parasitizing crab of the genus Portunidae.
Photo: Silliman University, Marine Laboratory

6. Fishes

Signs: Lacerations of the clam's mantle. They are mostly smaller than the ones caused by crabs. Constant harassment by fishes will leave the clam's shells partly closed.

Cause and course of event: Most fishes are active by day and the observant aquarist will soon pin the culprit. Not only fishes with a natural habit of eating mussels must be considered dangerous but other species as well whose behaviour may be changed under aquarium conditions, probably caused by dietary factors or just idle play.

Consequences: Mostly minor injuries on the siphonal mantle. If the fish can be removed, the prognosis is good in an otherwise healthy clam living in the right environment and healing will take place. A laceration of the mantle will be

Butterfly fishes, like this beautiful pair of Chaetodon lineolatus, should not be kept together with clams.

Anemone fishes can be kept together with giant clams.

Also the big "Jackfish", that we found in a shoal of estimately 800 individuals near Apo Islands, Philippines, the clams don't have to fear.

A big shoal of Plotosus anguillaris. About two to three thousand individuals of which just a few are showing in this photograph, were roaming around between semiadult T. gigas in a clam nursery.

closed and missing parts regenerated if not situated too close to the pallial line.

A difficult situation will arise if the fish identified as the culprit cannot be removed from the tank. This I experienced with one *Chelmon rostratus*. This species is known for its rugged behaviour against clams. This individual had lived peacefully amongst the clams and only after the introduction of several new tridacnas it turned to rip small pieces out of the mantles or gills and if the clam was on its side the fish entered the byssal orifice with it's pointed snout to get hold of soft tissue. This change in behaviour towards the clams was possibly triggered by a new key stimulus, probably a colour pattern of the mantle simulating other food items.

Similar behaviour can be observed in the cleaner fish *Labroides dimidiatus*. This popular and mostly harmless fish whose cleaning behaviour was first described by the enthusiastic diver and outstanding ethologist Irenäus Eibl-Eibesfeldt, can certainly prove lethal to a clam. This, however, seems to be the exception rather than the rule as most of these fishes live peacefully together with the pretty mussels. The reason for such a changed behaviour is not in the fish but the clam. Certain colour patterns of the siphonal mantle may mimic certain parasites taken as food by the fish thus triggering the behaviour. The fish tries to remove the "parasite" and the clam will close its shells. If these attacks are continually carried out, they will not only cause injuries to the mantle but the clam closing threshold will be lowered. The closed intervals will be lengthened and therefore the action of the symbiotic algae impaired. The clam will be made available for bacterial infections due to a low resistance and will finally succumb to the constant harassment.

Precaution and counteraction: The best precaution is to avoid fish which naturally feed on mussel. Therefore triggerfish and parrotfish are out of question but aquarists easily fall for labrids, which can grow quite large, or butterfly fishes or emperor fishes. Sometimes these fishes abruptly change their behaviour towards corals and mussels after having peacefully lived together with them for years.

One of the only labrids I used to tolerate in a clam aquarium are juveniles of *Coris gaimard,* as already mentioned. It is recommended as it helps to fight parasites. If the fish changes from juvenile to adult, however, not only the colour pattern is changed but the dentition becomes stronger, the behaviour becomes altogether more rugged and the appetite grows. Such individuals should be removed with the help of traps. J.Sprung (1996, pers. comm.) reports succes with *Macropharyngodon* and *Pseudocheilinus* (six line) wrasses.

Fossilized tridacnid clams. Above we see a fully stoned T. squamosa, an adult specimen with a shell length of 30 cm and the picture below shows a T. maxima.

6 Keeping Clams in the Aquarium

In an aquarium with a closed circulatory system, in contrast to tanks being constantly flooded with fresh seawater, only a relatively small number of reef animals can be kept. Many aquarists are keeping an astonishingly large number of animals in their tanks but in comparison to the natural condition in the reef, the selection is rather small.

The blue coloured specimen of T. crocea belong to the mostly favoured invertebrates in reef tanks.   *Photo: Svein A. Fosså*

In well managed reef tanks a high number of invertebrate species can flourish.   Photo: Klaus Jansen

With lime stone naturally appearing rock
formations can be arranged in the aquarium.

A T. derasa with beautiful patterns in a reef tank.
Photo: Svein A. Fosså

Clams in a private reef tank.
Photo: Klaus Jansen

Giant clams can be associated with most
invertebrates we find in reef tanks.
Photo: Klaus Jansen

Many of the conditions vital to coral reef animals in the aquarium must be artificially created and monitored. Water movement and water preparation cannot be accomplished without technical aid. Replenishment of evaporated water, trace elements, or salt concentration, to mention only a few, are vectors of a natural physical regulatory system, not present in our aquaria. The food has to be artificially added as well as removed. Even the saltwater, our pets natural environment, is mixed together from a long list of chemicals. The interaction of all these technical and chemical processes with the highly complicated biochemistry of living organisms is still so imperfect that only highly adaptable animals from the reef can be kept. In comparison to the coral reef with its millions of years of evolution behind it, our aquaria are about as perfect as an incubator in comparison to a mother's womb.

Therefore it is highly astonishing that many coral reef organisms do adapt to aquarium conditions, flourish and multiply, and reach rather methusalemic ages. Fish *(Acanthurus xanthopterus, Zebrasoma veliferum)* surviving in good health for 18 years, soft corals *(Sinularia asterolobata)* for 17 years are convincing examples, whereby the latter only succumbed when transferred to another tank. However, much has been accomplished in the last two decades through the enthusiasm and technical ingenuity of aquarists. Without them we would not stand where we are now. This should be highly appreciated.

Clams are relatively hardy and well suited for the saltwater aquarium. Many aquarists have kept these attractive molluscs for many years, recording their impressive growth. Slowly the shells grow and the mantle becomes mightier. Depending on the species, individuals

the size of a fist can reach a length of 25 cm or more in 6 or 7 years, *T. gigas* and *T. derasa* even much faster.

On the other hand it is not easy to explain the great popularity of clams. Certainly the sometimes indescribable colour combinations on the siphonal mantle, creating marvellous mosaic patterns partly account for that. Even in photographs taken with artificial light in the reef and showing the bewildering colours and patters of coral reef organisms, clams often stand out clearly with their metallic turquoise or blueish iridescent colour patterns.

Other characteristics may add to these attractions. Here, the hardiness in captivity is certainly an important factor. The clams do not need to be specially fed, they occupy little space and do not have an invasive growth pushing other organisms out of their niches. Their shells are a suitable habitat for many marine organisms. With their expansion of the mantle, its retraction when the shells are being closed and the slow reopening reminiscent of the opening of a blossom, they show more signs of life than most of the other invertebrates. All this must be taken into account when explaining their popularity. They are peaceful animals, in contrast to soft and stony corals which have evolved special strategies to conquer areas occupied by other organisms by pushing them aside and by doing so, depriving them of their habitat. None of this invasive behaviour will be observed in clams. To the contrary, the shadowed areas will provide a suitable habitat for sponges and other organisms. Under natural conditions, the invasive spread of corals makes sense, compensating for natural losses caused by predators. In the aquarist's tank, however, a flourishing organism with invasive strategies could dominate the whole community by pushing aside the co-inhabitants. A good example are the *Aiptasidae,* having a reputation of prolific growth with domination of all other organisms in the tank, leaving the aquarist with a monoculture. Many corals spread by harming their neighbours with their stinging tentacles, even strategies acting indirectly have been developed by releasing poison enclosed in mucus into the open water thus harming the other organisms and by this means conquering space and creating favoured conditions for their own survival. This competition with biochemical weapons in the confines of a closed system may lead to the unexplained degeneration of one group of organisms in an otherwise flourishing community.

Clams avoid this sort of competition. They do not produce poisonous substances and are themselves resistent to the nettles of corals, thus being also protected against the floating poison of the corals. Therefore, clams can be kept together with nearly all sedentary corals, probably a further reason for their popularity.

Similar to corals, which, in addition to the microplankton also use assimilation products of the symbiotic algae as nutrients, the feeding of clams is rather simple. They do not have to be fed directly like some corals from deeper reaches of the reef with its dim illumination, like species of the genus *Dendronephthya.* Such food would rather be an irritant, causing the clams to reject it because of its coarse structure. An indirect way of feeding clams with a plankton substitute is sufficient, similar to corals originating from the light-flooded upper reaches of the reef. However, the size of the particles must be adequate. Pulverized dry food is not recommended the particles being still too big. Liquid food like suspensions of yeast should be used. As already mentioned in the chapter about clam anatomy, the gills of the *Tridacnidae* are best suited to filter the so-called nanoplankton from the seawater.

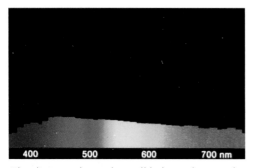

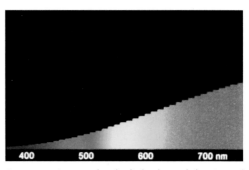

The diagram shows the well balanced spectral composition of the daylight. Fig. after Osram

In comparison to the daylight the red dominated light of a normal bulb. Fig. after Osram

Prerequisites for the Keeping of Clams.

Illumination

As already mentioned in previous chapters, clams have certain light requirements and they must be met under aquarium conditions. Not only the amount of light available but the physical composition of the light source is of utmost importance. The biophysical properties of the light are therefore more important than esthetic considerations. At this point we should explain the physical and physiological properties of light in more detail. Light, or what we

The appearance of a huge reef tank can be astonishing.   *Photo: Klaus Jansen*

A colony of Acropora with a diameter of about 15 cm after being newly introduced to the tank.
                    Photo: Rolf Hebbinghaus

The same colony after less than two years. The diameter is about 55 cm and the background shows several daughter colonies produced by this Acropora. Illumination: halide lamp, 10000 Lux.   *Photo: Rolf Hebbinghaus*

call light, is the effect of electromagnetic waves. There exists a broad spectrum of these waves depending on their length which is expressed as nanometer (nm). Only a small sector of light is recognized as such by the human eye, namely the range from 380 to 780 nm.

Above 780 nm we have infrared light which is used in medicine for therapies as it produces heat. A further rise in wave length will produce heat. At the opposite side of the scale, below 380 nm, we find ultraviolet light. This can cause acute or chronic damage to the

Acropora pulchra in the aquarium. The red marks show the colony size in 1991 (Löbbecke-Aquazoo).
                    Photo: Rolf Hebbinghaus

The documented growth of the Acropora colony shown left. The increase of coral surface is 1350 %. After two years the volume of the colony was 60 x 40 x 40 cm.
                    Photo: Rolf Hebbinghaus

human skin, leading to sunburn or even skin cancer. As already mentioned, the visible range of light lies in between the infrared and ultraviolet sector. The diagram shows that this sector is again divided into subsectors; the spectral colours. They span the range from violet (380 nm) to blue, green, yellow, orange, and red (780 nm). A rainbow impressively shows the sequence of the spectral colours. Daylight without a distinct colour represents a mixture of all these colours. The particular rays of one of the colour ranges can only illuminate objects reflecting these particular waves. Therefore red objects will only reflect the waves of the red sector thus appearing in the colour red. A green object cannot reflect red waves and will appear grey. Only the particular sector of the spectral range reflected by an object can be recognized by the human eye. What we call daylight is a homogenous mixture of all the waves of the spectral colours. This spectrum, however, will penetrate only the uppermost layers of the sea. There is a simple reason for that: the water acts as a spectral filter, eliminating certain wave lengths in relation to the height of the water column. If diving only to to a few meters depth, the red colour will be eliminated by the filtrating power of the water column. A red fish, for example, will show a bright red colouration near to the water surface but in a few meters depth the red will turn grey as the water will not contain the red sector of the spectrum and therefore no red rays will be reflected from the surface of the fish. The red colouration will have turned grey, camouflaging the animal against it's surroundings.

The same happens to other spectral colours. A fish coloured yellow at the surface will also turn grey but this will happen in a greater depth as the yellow range of rays can penetrate the water to a greater depth than the red sector of light. If we go deeper even the green colour will fade and everything will appear blue. The blue and violet part of the spectrum, however, penetrates to a great depth and thus everything really blue will appear in this colour. All the rest, that is the larger part of marine animals, will appear in inconspicuous grey and will thus be perfectly camouflaged.

Of course the whole system works the other way round as well and a fish with a pattern able to reflect the yellow part of the spectrum will appear bright yellow when rising to the upper reaches of the water column. Now a conspicuous pattern can assume a signal value, rejecting predators, facilitating schooling behaviour or starting reproductive behaviour. All these interpretations assume that fish are receptive to the same spectrum as the human eye. If future investigations will prove this not to be so, some of our interpretations of fish behaviour would be wrong.

For the sessile invertebrates of the reef with symbiotic algae, the physical properties of light and water are of utmost importance. They are completely adapted to the surrounding light conditions. The symbiotic algae employ pigments as aids for photosynthesis which are specialized to certain wave lengths, in other words certain colour sectors. These pigments are only able to absorb a certain sector of the spectrum of light which they use for photosynthesis.

The most important of these so-called assimilation pigments are the chlorophylls of which eight have been described up to now. The most frequent of these is chlorophyll A, active between 430 and 670 nm. This is a rather broad band of the spectrum. The most active parts in this band are the blue and orange ranges, the green-yellow ranges being less active. Other assimilation pigments are represented by carotine (a precursor of vitamin A) and carotinoids (substances chemically related to carotine) and are suited to utilize blue shortwave rays of light. These pigments are better suited for the deeper ranges of the reef with predominantly blue wave bands. For example, beta-carotine absorbs the range of 450 to 500 nm, representing the blue-green zone, and phycobilline the green-yellow-orange range from 500 to 650 nm. Therefore, they cover the range of light in which chlorophyll A is less active, thus complementing each other.

Marine algae normally indicate by their colour which assimilation pigments are utilized for photosynthesis. A greenish colouration hints to chlorophylls, red colour is indicative of carotine or carotinoids. Accordingly, different species of algae can be associated with certain depth ranges; blueish algae, mostly of a slow growth rate, are found in the deeper parts of the reef, the fast growing green algae populate the upper reaches. The different algae are adapted to the range of light waves present in their natural habitat and keeping them in captivity under a different range will prove difficult or even impossible. A Caulerpa algae will wither under pure blue light, because its assimilation pigments are not able to absorb it.

A similar situation we find in sessile invertebrates harboring symbiotic algae in their tissue. Most of the species occur in the sun flooded, shallow parts of the reef and are used to a broad spectrum of light. However, some of the organisms are specialised for greater depth and are utilizing auxiliary pigments

predominantly active in the blue range of light. Similar to algae, invertebrates occurring in greater depths can be identified by their colour. For instance, disc anemones show a red colouration if they harbour pigments of the carotine family. It goes without saying that the invertebrates must be kept under light conditions they are adapted to.

Further adaptations are necessary for survival in the coral reef. Invertebrates utilizing symbiotic algae in addition to filtering plankton, are autotroph (utilization of inorganic substances like carbonate, carbon dioxide, mineral salts: mode of nutrition in plants) as well as heterotroph (uptake of organic matter like protein, sugar etc., mode of nutrition in animals). This double barreled way of nutrition is widespread in invertebrates of the coral reef and some of the species have mastered this speciallisation to a high degree, switching from autotroph to heterotroph and vice versa, according to the prevailing requirements. This is why individuals of one species are able to survive in the sun flooded surface waters as well as in deeper zones of the reef which normally provide more planktonic particles. Chalker & Dunlap (1984) reported about the ability of the hermatypic coral *Acropora* sp. to keep up the same rate of photosynthesis over a depth range of one to 35 m. In one meter depth, 85 % of the light of the sun is available to the corals, dropping to only 8.5 % at 35 m. In greater depth, the coral can only exist without considerably improving photosynthesis by switching to heterotrophic carbon uptake, i.e. filtering more planktonic food.

The clams are able to vary the rate of photosynthesis according to requirements. This was found out by the Philippine biologist Suzanne Mingoa-

Young giant clams of the species Tridacna gigas.

Photo: Marine Laboratory of the Silliman University

Licuanan (Mingoa 1988). 45 juveniles of the species *Tridacna gigas* with a shell length of 12 to 20 mm were exposed to intensive sunlight. The same amount of animals was placed in the shadow with a reduction of sunlight to one tenth of its normal intensity. The animals were kept under these conditions for 53 days and then thoroughly investigated. The outcome was very interesting. The young clams were able to cope with a reduced light intensity by doubling the amount of chlorophyll A in their symbiotic algae. As a consequence, the algae living in the clams exposed to the full intensity of the sunlight needed more light to produce a certain amount of oxygen than the control group from the shadowed tanks.

This means that young clams are able to govern the amount of photosynthesis in the algae according to the amount of light available at a certain site. However, it is not known if adult clams also process this ability, for example, if placed in an aquarium with a drastic change in light intensity. But this is only one way to cope with a change in light intensity. The number of the photosynthetic units (PSU) can be altered or, as already described in the chapter on clam anatomy, the number of symbiotic algae in the siphonal mantle can be altered. It is not yet clear, however, if these mechanisms are present to the same amount in all species of the *Tridacnidae,* or if species specific differences exist, enabling one species to cope with changes in light intensity better than others. Yonge (1975) suggested that the enormous growth rate seen in the *Tridacnidae* can be attributed to the photosynthetic activity in the siphonal lobe, rendering the animals relatively independent from food suspended in the surrounding sea water. However, this would mean that clam species growing to a large size would be more dependent of a high light intensity than smaller individuals. In case of diminished sun light, small species of clams, like *T. crocea,* should be able to compensate for nutritional losses by raising the amount heterotrophic nutrition. Therefore, smaller species like *T. crocea* should penetrate to greater depths with little light for photosynthesis, than larger species like *T. gigas* which should be restricted to shallower water with enough light intensity to keep their symbiotic algae flourishing. However, as a matter of fact, the contrary is the case with the smaller species being found near the surface of the reef indicating a need for higher light intensities and a supposedly smaller adaptability to vary the amount of autotroph and heterotroph nutrition. Huge specimens of *T. gigas* can be found in 20 m depth where a *T. crocea* would hardly have any chance to survive. I suspect that the larger species are better adapted to greater depth with its diminished light intensity than smaller clams. It is known that the smaller an animal, the higher is its metabolic rate. That means that a juvenile specimen of a genus will have a relatively higher energy turnover than an adult individual. It is also possible that

The right spectral composition of the illumination is one of the most important prerequisites for a reef tank.   *Photo: Klaus Jansen*

the larger species like *T. gigas* have a higher ability for heterotrophic nutrition than the small *T. crocea*. On the other hand, light protection mechanisms seem to be better developed in the small and sun-seeking *T. crocea*. "Central bleaching", a phenomenon described in the chapter on clam diseases and attributed to sudden high intensity of ultra violet light was only found in larger species like *T. squamosa* or *T. gigas*, but never in the small *T. crocea*. That means that, concerning light intensity, no matter if natural or artificial, the larger species of clams must be handled with greater care than the small *T. crocea*. Up to now this is all conjecture without prove but the observations point to an interspecific difference in adaptability to light intensity.

Keeping the *Tridacnidae* in aquaria, you will nearly always need artificial light source. Unrestricted sunlight will propagate too much heat and will be only feasible if the aquarium is connected to a larger tank working as a cooler by constantly exchanging warmer for cooler water. A pump failure in such an arrangement, however, can have a disastrous effect by dramatically raising the temperature in the sun exposed tank.

If artificial light is to be used, not only the intensity but the spectral composition is important.

The total sum of the spectral colours of a light source making up the colour of the light or the light temperature, measured in Kelvin. The blue, short wave sector of the spectrum has a high colour temperature and is high on the Kelvin scale and the red, long wave sector with a low colour temperature is low on the Kelvin scale. Sunlight has about 5800 Kelvin, with meteorologic oscillations between 3000 to 30000 Kelvin, but an ordinary light bulb will only reach around 2500 Kelvin. The Kelvin scale alows for a rough estimation of the spectral components. It should be obvious that a normal light bulb with its high portion of red differs considerably from natural day light and is thus highly unsuited for our clam aquarium.

However, as mentioned, the Kelvin scale allows only for a rough estimation of a light source. For example, a high portion of red light will lower the colour temperature, can be compensated by a high portion of blue light with a high colour temperature. Mixed light will not change its colour temperature if both spectral components will be intensified simultaneously. The rest of the spectral colours, however, would be diminished and the light source would primarily emit red and blue. An animal whose symbiotic algae are primarily active in the green-yellow range could hardly survive under such light conditions, even if one would considerably raise the overall light intensity. On the other hand, certain primitive algae flourish very well in the red spectrum of light and would turn out to be real pests which could hardly be controlled. The Kelvin value, however, of such a light source could be very similar to natural daylight.

A much better and more precise way to present the characteristics of a light source can be obtained by showing the spectral components in a histogram. Such a presentation of the whole range of the visible light from 380 to 780 nm gives a good optical impression of the spectral parts present and the imbalance of an artificial light source in comparison to natural sunlight. Natural light is composed of the different spectral components in a rather uniform way but this is rarely the case in artificial light. This can clearly be

130

demonstrated in the histogram of a fluorescent tube of the daylight type Lumilux 11.

The short wave blue light is concentrated between 430 and 480 nm, the long wave orange light around 610 nm and a strong concentration of the green yellow range is found around 545 nm, and still the Kelvin value is very similar to natural daylight. These irregularities in the quantitative distribution of the different spectral components are not without impact on biochemical processes. These effects can become more serious in older fluorescent tubes with considerably reduced short wave blue parts with consecutive raise in the relative contents of long wave red spectral components. This is not without effect on the algal photosynthesis. Under such conditions certain symbiotic algae will wither while undesirable microalgae like Spirulina will flourish. A constant quality of the light source in an aquarium can only be obtained by replacing the tubes after six to eight months, preferably in a consecutive way to keep the mixed light in the tank constant.

As "minimal light source" for an aquarium populated with invertebrates, fluorescent tubes of the different colour types may suffice. I know several tanks with excellent growth of coral under such conditions. My own experiences are also good. However, for the well-being of clams of the genus *Tridacnidae* such a light source may not provide the necessary light intensity. Intensity could be enhanced by adding more tubes over the tank, but the light produced by tubes would not penetrate deep enough to satisfy the requirements of the clams on the bottom of a deep tank. As the main light source, the only suitable means are metal halide lamps, which are now widely used in reef aquaria. Fluorescent tubes are only used as dimmers before and after the main light source is switched on or off. However, many aquarists use them as a corrective, for example to enhance the blue spectrum.

Mercury lamps, widely used in freshwater aquaria, should not be employed when keeping giant clams because of their spectral composition.

Daylight type metal halide lamps with Kelvin values of between 5000 and 6000 are suitable for our purposes. These lamps, being on the marked from several producers, are normally designated "D" indicating daylight. The Kelvin value, however, will only give a rough estimate of the spectral components, as outlined above, and some lamps will change the composition over the time used, mostly in the range of several months and the original specification will at best be of statistical value only.

I recommend metal halide "TS" lamps of 250 Watt (Osram TS 250W/D) and tube shaped lamps with the socket E 27 or E40 (Osram T 250W/D). I also used halide lamps with 400, 1000, and 2000 Watt with no apparent negative effect on the clams, traced back to the light source. These stronger lamps should be used over deeper tanks and with a greater distance to the water surface. 250 Watt are used for tanks 60 cm high, 400 Watt with tanks of 80, and 1000 Watt with tanks of 100 cm water column. Lamps of 2000 Watt should only be used on tanks more than one meter high with a safety distance of at least 70 cm. Such lamps have an enormous light output. A 250 Watt lamp will produce 19000 Lumen, a 400 Watt lamp 33000 Lumen, and a lamp of the same type with 1000 Watt will produce already 80000 Lumen. A 2000 Watt lamp, however, has an output of an

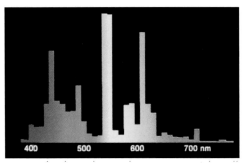

The spectral composition of a fluorescent lamp of daylight type (Osram Lumilux 11). Fig. after Osram

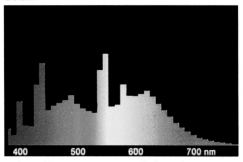

The spectral composition of the daylight fluorescent lamp Osram Biolux (Type 72) is more well balanced, but the light density is smaller. Fig. after Osram

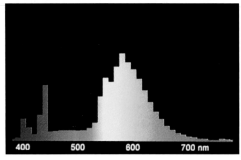

The spectral composition of a fluorescent lamp of warm white type. The light density is very high, but the light is dominated by red/orange spectral parts which makes it unsuited for giant clams. Fig. after Osram

Halide lamps of the manufacturer Sill, to be used with Osram halide bulbs HQI T 250 W/D or 400 W/D.

Photo: Sill

Halide lamp of the same manufacturer to be used with Osram bulbs HQI T 1000 W/D
Photo: Sill

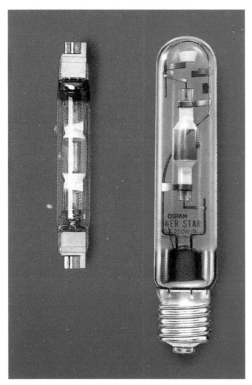

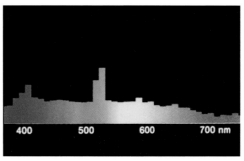

The spectral composition of Osram HQI bulbs of daylight type is well balanced and has similarity to natural daylight. Fig. after Osram

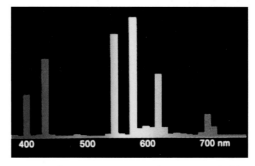

Mercury lamps Osram HQL show a remarkably unbalanced spectral composition. Fig. after Osram

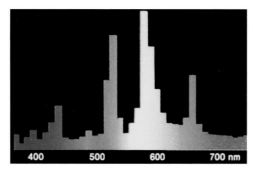

Halide lamps of warm white type (Osram HQI WDL) have a strong green/yellow and red radiation. Fig. after Osram

Osram HQI bulbs

astounding 170000 Lumen. Even in light- hungry animals, such a light intensity could be harmful. Therefore, the light intensity should stand in a reasonable relation to the height of the tank.

For some time now, daylight type lamps of 150 Watt are available. Seawater aquarists have long awaited this type. It replaces the till now widely used NDL type (neutral white de luxe) which, with its warmer, yellow biased colour and high ultra violet content, used to cause problems. These problems have now been rectified but with its lower light output it is of only marginal value for the *Tridacnidae* and should only be used for tanks of up to 50 cm height and the clams should not be placed in the lower third of the aquarium. As a rule, however, only daylight type lamps should be used. Other types like NDL (neutral white) or WDL are not recommended. Colder light temperatures, however, can be useful and lately metal halide lamps with a stronger blue spectrum appeared on the market. These lamps with Kelvin values of about 10000, produce light conditions very similar to those found in the coral reef in several meters depth. A tank illuminated with the new type of metal halide lamps does seem to be less bright in comparison to the conventional halide source but this is due to its lower green-yellow spectral components comprising the range around 550 nm. This range appears especially bright to the human eye. Such a lamp with about 10000 Kelvin can be used as the sole light source of an aquarium. Certainly the future will bring further improvements. Already on the market is a device for varying a metal halide lamp's light output by reducing it by 60 %. This device, manufactured by Sylvania is still expensive due to low production numbers. It allows to simulate a near perfect dynamic of the light conditions and is certainly a further step in the right direction. It is interesting to note that this type designated ILS 2000 emits relatively more blue spectral parts in the dimmed phase, thus producing a very pleasing effect.

In case of daylight lamps with 5000 to 6000 Kelvin, it is advisable to correct the spectrum by adding blue fluorescent tubes. Such an arrangement will not eliminate the red and yellow spectral sectors but the animals will be kept under a mixed light with the correct blue spectrum. There are several blue tubes on the market which can be used to this effect (Osram Type L 67, Philips Type TLD 18). The combination of a metal halide lamp and a blue

fluorescent tube, years ago propagated by Peter Wilkens, proved to be very beneficial when keeping invertebrates. Being used to the full spectrum of light, the human eye has to adapt somewhat to this type of illumination. However, we should keep in mind that the light conditions must satisfy the physiological needs of the animals rather than our aesthetic feelings. The only drawback giving a rather unnatural impression is the sudden change in light colour when the lamps are switched on or off. For instance, if a metal halide lamp with about 10000 Kelvin is switched off, a very sudden rise to about 25000 Kelvin will occur. Of course, such an instant change in light intensity will not happen under natural conditions and we should keep in mind that our animals originate either from the sun flooded parts of the reef or from deeper ranges with a predominance of the blue spectrum of light and are adapted respectively. It does not seem advisable to stress the animals adaptability to the limit and probably we find here the reason for difficulties we encounter by trying to keep certain groups of animals in aquaria.

For optimal artificial illumination of the clams, I advise the addition of a daylight type fluorescent tube like Osram Lumilux 11, 12 or Biolux 72 in combination with metal halide lamps and blue-light-type tubes. Such an arrangement will not noticeably change the mixed light of the combined light sources, the metal halide lamp and the additional light having similar mixed light. The overall impression will be similar to natural light and the organisms will appear in their natural colours. However, some aquarists appreciate the effect of blue light producing marvellous green iridescent colouration in certain animals. Especially the blue and turquoise shades of the clams mantles will become enhanced under blue light.

In general the bright colouration of the clams is more obvious when observed from above than through the sides of an aquarium. The explanation is simple. What we perceive as colour in the clam's mantle is in reality a so-called "physical colour", an effect of the optical refraction of light. In other words, the colours we see on the mantle are in reality illusionary like the colours of a rainbow. If sunlight hits the raindrop under a certain angle, only a fraction of the whole spectrum will be perceived and the rest filtered out. If the raindrop falls, the angle will change and a different sector of the spectrum will be reflected. If it falls further, it will reach a point where no reflection takes place and the observer will see the normal mixed light.

A very similar mechanism is responsible for the colours emitted by the clams' mantles. The highest degree of refraction, however, takes place when the clam is observed from the same direction from which the light comes. The most advantageous way to observe these animals is therefore from above in the direction of the light source. Being aquainted with these facts will help when taking photographs.

After this distraction let us go back to the illumination of our tanks. I recommend a daily light period of about 8 hours when using metal halide bulbs. Using a weaker light source, the diminished light intensity could be compensated for by prolongation of the illumination period but this is only possible within reasonable limits. Fluorescent tubes employed to correct the light temperature can be left on for about some hours before and after the metal halide lamps are switched on or off, producing the effect of dawn and

dusk. The light period must not be stretched to its limit in the assumption to enhance the photosynthetic activity of the symbiotic algae. It is true that the photosynthesis is a light-dependent process, however, it is divided into two phases. During the 'light reaction' photosynthetic pigments (chlorophylls, carotinoids, phyllobillines etc.) will transform light energy into chemical energy through complicated biochemical pathways. At the same time, more or less as a byproduct, oxygen is released. During the "dark phase", only possible outside the illumination period, carbondioxide is transformed to glucose.

At least ten hours of total darkness should be allowed for this chemical reaction in the algae. During this period also the fluorescent tubes employed for colour correction should be switched off.

Metal halide lamps must be handled with the necessary care and the installation should be carried out by professional electricians. Safety distances must be observed and ultraviolet filters must be installed. Nearly all metal halide lamps emit, apart from the visible spectrum, non-visible short waved ultraviolet light of A, B, and C characteristics. They are harmful to human beings and must be eliminated with the help of silicate glass filters. Metal halide lamps must never be used without this precaution. A safety distance of at least 30 to 40 cm between filter and water surface is also important to protect our pets from an overdose of harmful rays. Bleaching and loss of symbiotic algae would follow because the irradiation would considerably shorten the algae's life cycle. Precautions are also necessary when exchanging metal halide lamps to avoid this "photo oxidation". New metal

The author's Nephtheidae culturing tank from 1988, illuminated with two Osram HQI TS 250 W/D lamps. The giant clam species shown are T. crocea (left) and two T. maxima (right).

The view of a huge reef tank, illuminated with Sill lamps 1000 and 2000 Watt and fluorescent lamps of blue type.
Photo: Klaus Jansen

The 1000 liter reef tank of Robert Wong, Manila with marvellous growth of many stony coral species.

Technical control and protein skimmer of the tank shown above.

halide lamps are very powerful and this gain in intensity should be checked by either a greater distance between lamp and water surface or the use of opaque glass plates.

Skimming

The coral reef is a closed biochemical system. All substances excreted as waste by one organism are utilized by an other as valuable raw material. Therefore, 'waste' is a foreign word in the reef. Overproduction of one substance will nearly always be followed by a rise in the population of the organisms utilizing this particular substance. Such control circuits prevent the imbalance of the ecosystem.

The aquarium, however, represents only a small fraction of the whole coral reef system with only a few of the regulatory systems present. Especially the transformation of organic into inorganic matter, known as "mineralization" being the opposite to photosynthesis, is only possible in small scale in the aquarium. With little waste output, it is possible to establish such a system in an aquarium. If a 500 l tank is populated by three small fishes and about 10 corals with plenty of space for higher algae and the development of a micro fauna, the fish do not have to be fed. The production of organic waste will not be greater than the mineralization by bacteria and vice versa the mineralization will not supersede the activity of the photosynthesis where inorganic matter will be transformed into organic substances like protein and carbohydrates. In this case the system is balanced and will stay stable over a considerable period of time, provided that all the other environmental conditions like illumination, temperature, and salinity are stable.

In most cases, the reality looks different. Often the aquaria are overpopulated with fishes causing considerable organic waste in the form of excess food and feces. Mineralizing bacteria can not cope with such a workload. Without technical aid, toxic substances like ammonium would accumulate, derailing the system.

Skimming is a physical process, constantly eliminating toxic waste from the system and thus cleaning it in a continuous way. Filter systems using foam or similar material are only holding the undesirable matter without eliminating it from the system entirely.

Skimming has also a positive effect on the gaseous exchange by eliminating noxious gases and enhancing the concentration of beneficial ones, mainly oxygen.

The skimming process consists in mixing diminutive air bubbles with aquarium water causing constant friction between the two components. This friction causes the development of an electric load on the surface of the air bubble which attracts certain ions, mostly carbohydrates.

The important advantage of the system is attributed to the fact that the carbohydrates thus bind substances dissolved in the water. These substances adhere to the air bubbles surface and can be eliminated from the system. In other words, the carbohydrate functions as a vehicle for dissolved substances contained in the seawater.

Technically the system is very simple. Large amounts of small air bubbles are produced by aerating a tube filled with aquarium water. To prevent the bubbles from immediately reaching the surface, a constant flow of aquarium water against the direction of the bubbles is produced by a pump. In this way, the reaction time between water and air is greatly prolonged and therefore more substances can be loaded onto the bubbles. The effectiveness of the filter is therefore not only determined by the amount of water running through the tube but also the time of contact between air and waste loaded water.

The surface of the system will carry the bubbles, saturated with carbohydrates. They will not burst. Further down, pure air bubbles without a coat of carbohydrates are found. They will eventually burst and the free air will press the firm foam loaded with waste to the surface where it will be collected in a special container.

Of course such a system can not distinguish between harmful and bene-ficial substances necessary for the well-being of the fauna and flora of our tank. However, this is no argument against the use of the skimming principle as the substances eliminated from the water can easily be replenished.

The formation of the foam is dependent on the surface tension of the water. Substances reducing this tension can abruptly terminate the foam formation. Feeding fish with fat-containing food, for instance, can immediately cause the foam to collapse. It can take several hours for the foam to become re-established. It is important to keep the system clean by removing the foam regularly and cleaning the tube thus eliminating fatty residues which can inhibit foam formation.

Biological Filtration Systems

Two thirds of the waste products in an aquarium are surface-active and can be removed by skimming. The remaining third, however, has to be metabolized by bacteria. In other words, it has to be mineralized. A precondition for the existence of the bacteria is a firm substrate as they do not live in open water. Therefore, bacteria, mainly *Nitrosomas* and *Nitrobacter* species, are necessary for purification processes, even in the presence of a powerful skimmer. Normally, surfaces provided by decorative rockwork suffice and in the case of porous material it will be also populated under the surface by bacteria. Nitrate synthesizing bacteria will populate the oxygen-rich surface of rocks, turning ammonia into toxic nitrite and in a further step into the relatively harmless nitrate. The oxygen necessary for this process comes from the surrounding water. During this process which is only roughly outlined here, other metabolites like nitric acid and phosphoric acid are produced. To prevent an accumulation of these sub-stances, other creatures are necessary, for example anaerobic bacteria. In contrast to the before mentioned species, anaerobic bacteria need an environment free of oxygen. However, they still need oxygen for the their metabolism but do not utilize oxygen from the surrounding water but oxygen trapped in waste products where it became incorporated during the process of oxydization. Taking out the oxygen they reduce nitrate back to nitrite. A further step will then reduce nitrite to gaseous nitrogen which can be eliminat-ed through the action of the skimmer. In reality, the process is much more com-plicated and only roughly out-lined here.

All these steps take place inside the porous lime stone. The outer regions are

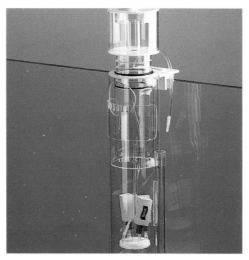

Protein skimmer from KNOP-Aquarientechnik
(type Maxi 100) for aquariums containing 200
to 1000 liters.

External skimmers from Royal Exclusiv-
Aquarienmöbel und Anlagenbau.
Photo: Klaus Jansen

External skimmer from KNOP-Aquarientechnik
(type Mega, different sizes serving tank volumes
from 500 up to 15000 liters).

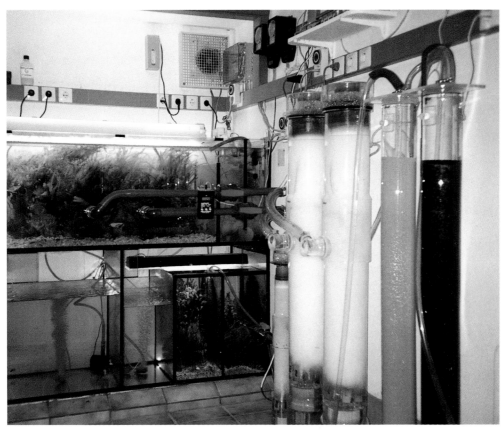

Water replenishment and protein skimming in a huge experimental reef tank of KNOP-Aquarientechnik.

The scum pot of Mega 200 after a few days of skimming action.

responsible for the oxidative and the inner regions for the reductive processes, the former turning ammonia into nitrate and the latter metabolizing nitrate further. Such a piece of rock, ideally from the coral reef, represents a perfect biological filter and will provide our aquarium with a well balanced nitrogen metabolism.

Numerous attempts have been made to reproduce this system by developing technical filters simulating the natural situation. In the case of nitrate build up they have met with great success and a large number of bio-filter systems are available using the action of *Nitrosomonas* and *Nitrobacter* species. Unfortunately, this is not the case with the nitrate reduction. Most of the artificial systems are rather 'oxydation heavy', producing constantly nitrate and other metabolites which are then accumulated reaching concentrations toxic for fishes and invertebrates. Corals and clams are more sensitive to such waste accumulations than most are most fishes and therefore will soon show signs of discomfort. Trickle filters or wet-dry filters are able to turn organic material into nitrate but are not able to metabolize the accumulated nitrate further. Submerged bio-filters, mostly filled with porous limestone fragments, have a tendency to clog-up with debris and thus become ineffective. Such a system is hardly controllable and if the nitrates are used up and are not replaced, the anaerobic bacteria will necessarily stop metabolizing nitrate and will switch to the so-called "sulphide metabolism". In this case sulphate will be reduced to sulphide with the production of free oxygen which will then be utilized by the bacteria. The end product of this reduction would be hydrogen sulphide with the characteristic smell of foul eggs. Such debris with its black

colouration can cause extensive damage when suspended in the tank.

A useful supplement to the action of living reef-rocks is the activated carbon propagated by Peter Wilkens under the trade name HydroCarbon. Even after saturation with toxic waste it will work as a biofilter substrate. Inside the carbon corpuscles we find a redox potential of -50 to -200 millivolts, being the range of the bacterial nitrate reduction potential. Due to the size of the carbon particles a redox potential lower than -200 millivolts can not develope, necessary for the sulphate reduction. Numerous investigations have shown that this material effectively allows the control of the nitrate contents of the seawater.

Bacterial metabolism, or mineralization, does not only lead to nitrate production but to the production of a whole range of other products as well and it is not advisable to lean completely on bacterial action as filtration. Let us remember that the metabolic pathway includes plants animals and bacteria. The first step consists of the "autotroph" organisms, the plants and the symbiotic algae living in certain animals; by the action of light organic substances like carbohydrates and protein are synthesized from non-organic matter. Those products are now used by the 'heterotrophic' inhabitants of the reef, comprising the second step; the carbohydrates are broken down into their respective components and the protein will be turned into polypeptides and further into amino acids, the smallest compound of protein. These components are utilized by the animals to synthesize products to their own specification which results in secretion of their own organic waste products. now we come to the third step; bacteria metabolize organic waste into inorganic material thus closing the metabolic circle. This system only works if the different components are present and active in a balanced quantity, producing a closed ecosystem. This holds true for a natural water body, like a pond, a lake, or a coral reef but also for our aquarium. If we introduce organic material into such a system, metabolites can accumulate at any given step in the system and the balance will be impaired. In the presence of many heterotrophic animals, mostly fishes, the food resources will be utilized resulting in organic waste products. In the presence of sufficient numbers of bacterial colonies, the organic waste will be mineralized. This, however, is the point where the system is easily thrown out of balance because of the lack of sufficient autotrophic organisms turning organic matter back into organic substances.

This over-fertilization, well known from inland waters as well as the North Sea, produces an ecological niche for lower algae with their enormous growth potential leading to explosive rise in their populations. Filamentous algae will spread over the entire decorative rocks in an aquarium and diatoms will produce gigantic algal carpets on the surface of the sea, both being the consequences of over-fertilization. This is nature's way to cope with an imbalanced system; the autotrophic organisms are now in action, metabolizing inorganic matter into organic substances but often to the detriment of other organisms by destroying their environment.

To avoid such conditions in our little ecosystem we call 'aquarium', an overdose of organic matter must be removed before the bacteria begin with the step three in our scheme; the mineralization. Therefore, it is obvious that, apart from efficient biological filtration, the skimming process is of utmost importance if a balanced state is to be kept over a longer period of time.

However, if we do not keep in an aquarium the usual soft or stony corals but mainly clams, the metabolism of nitrogen, nitrite, and nitrate must be looked at from a different angle. Basically, we find the same situation with waste production due to food residues and excrements of the fishes, but they are taken up by the clams as food for the symbiotic algae in considerable amounts. Contrary to the usual attempts to keep the nutritional contents of the sea water as low as possible because delicate stony as well as some soft corals will show better growth under such conditions, dense populations of clams need higher concentrations of organic nitrogen compounds to satisfy the clams metabolic needs. Per surface area, clams house about ten times more symbiotic algae than corals (C. Belda, pers. com. 1994). This drastic raise in the number of algae, made possible through the specialised symbiont channel system of the mantle, raises the output of the photosynthesis but also requires a higher amount of nutrition. We must keep in mind that organic nitrogen compounds as well as phosphate and nitrate, normally considered evil by the reef hobbyist aquarist, can prove beneficial when keeping clams. As in other things, the right amount is of utmost importance. Every plant, or in our case symbiotic algae, needs a certain amount of organic nitrogen without which it is not viable. Only an undue raise in the concentration of these compounds will produce an adverse effect.

A few small clams in a not overpopulated tank will not reduce the nitrate concentration. If, however, the

Very high reef tanks often provide a very natural appearance. Photo: Klaus Jansen

A beautifully coloured Tridacna maxima.
Photo: Svein A. Fosså

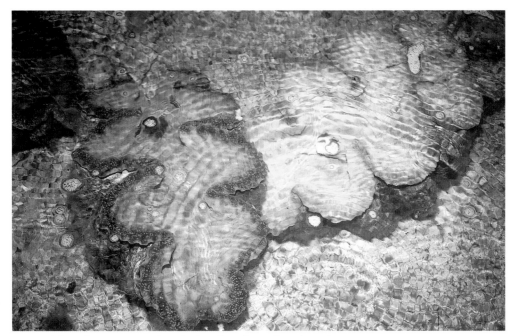
Semi-adult Tridacna gigas with typical syphonal mantle colouration.

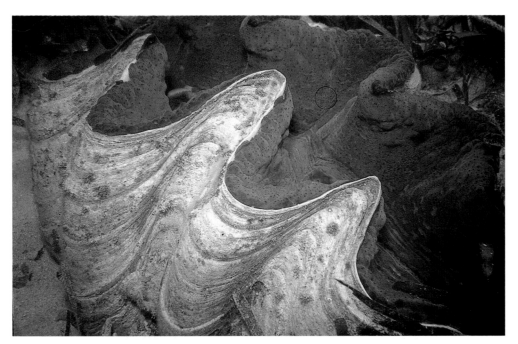
Adult specimen of Tridacna gigas provide an unforgettable impression to the diver.

| NO Ammonium | $\Rightarrow$ | NO_2 Nitrite | $\Rightarrow$ | NO_3 Nitrate | $\Rightarrow$ | NO_2 Nitrite | $\Rightarrow$ | NO/N_2O Nitric Oxide | $\Rightarrow$ | N_2O Nitrous Oxide |

The steps of the nitrification cycle.

tank is mainly populated by clams, the amount of measurable organic nitrogen will be considerably reduced. Unfortunately, it is not possible to give any rule of thumb as to the amount of clams per liter aquarium water. Conditions and circumstances are too varied to allow for such an estimation but the aquarist is well advised to check the nitrogen and phosphate contents of the water in regular intervals. It should be clear that an extremely low nitrogen or phosphate concentration must not be regarded as a positive sign in the clam aquarium. Seemingly inexplicable death or sudden bleaching should direct ones attention to the possibility of low nitrogen levels. In the chapter about clam diseases we have thoroughly discussed the conditions caused by a deficiency in organic nitrogen compounds and its remedies. It should be obvious that the keeping of clams with their high requirement for organic nitrogen opens new ways for controlling these compounds.

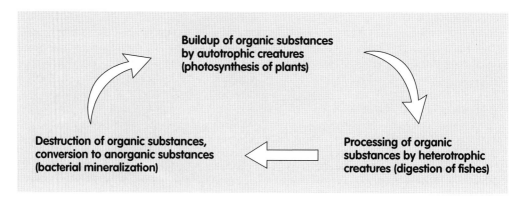

Buildup and destruction of organic matters.

Currents

Like all the sessile invertebrates, the members of the family *Tridacnidae* are not able to search for food or to move to other parts of the reef which more favourable water conditions. Therefore they depend on a certain amount of water movement; the water carries the suspended food particles which form an important part of their nutrition. In addition the current provides the animal with oxygen rich water and also carries away carbon dioxide produced by the clam. Minimal water movement is therefore vital for these sessile molluscs.

However, in contrast to a number of soft corals in the shallow brightly illuminated litoral, some of which have adapted to extreme currents, our clams prefer rather soft water movements. Corals of several species might get rid of their body secretions by strong currents, yet for our clams those turbulent waters would be rather disturbing. Firstly, they would disrupt the slow transport of water through the chamber system of the clam's body, secondly, it would hardly be of any help to the symbiotic algae if the syphonal lobe would flutter before the current.

Thus, when we keep clams in aquaria, we have to imitate the natural currents in the sea that evolve from tides and temperature differences, just by providing slow water movements. While calculating the pump strength in common invertebrate aquaria such that each litre in the tank is pumped up to ten times per hour - for example in a 500 litre tank two pumps each delivering 2500 litres per hour - we have to do with much less performance when keeping clams. Half the amount of this rule of thumb, that is five times the aquarium volume in litres as conveying capacity per hour, seems reasonable to

me. However, this is just a raw estimate; the form of the tank and its equipment with rocks has to be considered, as well as the diameter of the pump's discharge pipe. Thin pipes will produce a thin yet strong jet which causes unnatural effects at its end point. We therefore suggest wider pipe diameters which deliver more water without some individuals being exposed to dramatic turbulences. Similar to tank lighting, one has to bare in mind that the water movement does not serve aesthetic purposes, but has to fulfill the animals' needs. Therefore we have to provide a widely distributed soft current without any jet and its shearing effects.

If pumps with an interval pulse control are used, where the discharge capacity between two values is variable, the maximum capacity of the pump may be chosen higher. In this case, we should still make sure that the clams are placed in the aquarium in such a manner, that the syphonal lobe is not tipped up by the current. Although this is completely harmless when occurring occasionally, however, this kind of disruption can curtail the well-being of our beautiful molluscs significantly, when happening regularly.

Water

Natural seawater contains all known elements in a dissolved form, some only in traces (trace elements), others in larger quantities (main elements). The importance of some of these elements for the marine fauna and flora is well known since long, especially main elements like sodium, potassium and calcium and some trace elements like strontium, iodine, and iron. However, no research has been carried out concerning the importance of most of the other elements. Presumably they do play a role in biochemical pathways but up to now nothing is known.

In this book only the elements important in the keeping of clams will be considered. The more interested reader is referred to the literature given at the end of the book.

Salinity

The most important component of seawater is sodium chloride with a high concentration when compared to the other constituents. With a concentration of about 35 ppt (parts per thousand) it is similar to the one found in our blood, emphasizing our descent from marine life. 35 ppt stands for 35 grams of sodium chloride dissolved in one liter water. Dissolving 20 kg in 600 l water means 33.3 gr salt per liter and roughly reflects the salt concentration of the oceans.

An extraordinary beautiful specimen of Tridacna maxima, surrounded by some herbivorous snails.
Photo: Rolf Hebbinghaus

▲
The refractometer enables precise measuring of the water salinity.

▶

The "China Clam" Hippopus porcellanus.
Photo: Marine Lab, Silliman University.

The largest Hippopus hippopus in the world in its original exhibition shelf (UPMSI).
▼

Largest *Hippopus hippopus* in the world.

Denominating the sodium chloride concentration in weight unit ppt is very precise but has a serious drawback for the aquarist as it is difficult to measure. To measure the salt contents of the water we utilize the fact that dissolved salt makes water heavier. If one liter of pure water weighs 1000 gr we have to add 35 gr after having dissolved 35 gr of salt. Every milliliter of water is now heavier than before dissolving the salt. If we put an egg into pure water it will sink because it is heavier than the water displaced by its volume. If put into a concentrated solution of salt, however, it will float because it is lighter than the displaced water. To get an exact measurement of the salt concentration one could theoretically measure how high the egg floats in the water.

Exactly this principle is used when we measure the salt concentration with the help of a density hydrometer. A hollow glass cylinder of known weight is put into distilled water and the height of the flotation is marked. After this calibration, a scale is added which allows an exact comparison of a particular liquid's weight to that of distilled water. The higher the salt concentration, the higher the specific gravity of the solution and the higher the density hydrometer will float.

However, water will expand its volume with higher temperatures. In other words, one liter of water of 10 degrees Celsius will contain less molecules than one liter at a temperature of 90 degrees Celsius. Consequently all measurements must be carried out at the same temperature at which the instrument was calibrated. This temperature, normally, 25 degrees Celsius, is indicated on the scale of the density hydrometer.

During scientific investigations on marine organisms a much more precise instrument is used to measure the salt concentration; the refractometer. The use of this precision instrument is even more simple than the density hydrometer but unfortunately it is very expensive and not in the reach of every body. Nevertheless, if one attempts to breed coral reef fishes or invertebrates in aquaria, the purchase of such an instrument is highly recommended. The small inaccuracy of the density hydrometer may be of little importance in the management of a normal seawater aquarium but if attempting breeding marine animals a refractometer becomes a necessity. Of course, several aquarists could buy an instrument for communal use.

The principle of the refractometer is simple and not only used to measure salt concentrations but also, for instance, solutions of sugar. It is an optical instrument, similar to a monocle, and measures the refraction index of a given liquid. It does not contain any moving parts and is thus free of wear. The refraction index of a liquid is proportional to the amount of dissolved matter contained in the liquid. This fact is the base of the refractometer. The higher the salt concentration, the higher the refraction index. A small experiment will explain the principle. If a rod is hold halfways into an aquarium and looked at from above, it will appear bent where it enters the water. This is because the refraction index of water is different from that of air. The higher the salt concentration, the more the rod will appear bent. Even if we look into the tank through the front glass, the effect can be recognized and the submerged rod will appear nearer to the observer than the non-submerged part. The same phenomenon is responsible for the seemingly smaller depth of our tanks when filled with seawater and, if diving in the reef, fish seem to be larger than they actually are. This effect also impedes the estimation of distances when taking pictures under water.

If a drop of the fluid to be measured is put onto the measuring plate, the salt concentration in ppt and the specific gravity in kg can be read on a scale by looking through the refractometer. Highly sophisticated instruments even compensate for the temperature at the time of measurement. Only refractometers calibrated for salt solutions, however, can be used. Instruments for the measurement of sugar solutions have calibrations different from the above.

The specific gravity of one liter of water with a salt concentration of 33 ppt is 1.022 kg. If the salt concentration is 36 ppt, the specific weight is 1.024 kg. The optimal range for keeping invertebrates is between 1.020 and 1.024 kg specific gravity. This is also the range suitable for giant clams, despite the large variation of salt concentration found worldwide. The only exception is the Red Sea with a concentration of 40 ppt and clams as well as other invertebrates originating from the Red Sea must be kept at a higher salt contents. This becomes especially important when keeping Xeniids from the Red Sea as they react promptly with degeneration if the salt concentration will be drastically diminished. If we are keeping clams from the Red Sea together with individuals from the Indo-Pacific or other tropical regions, the salt concentration should never sink below 33 ppt (1.022 spec. weight).

The salt concentration in the tank should be kept as constant as possible though some invertebrates are subjected to considerable changes in their natural habitat, especially when originating from shallow water in areas of heavy tropical rains. However, not all animals of our aquaria do have the same range of tolerance when salt concentration concerns and the impact of changing salt concentrations on the bacteria must also be taken into consideration. Therefore, daily measurements with a replenishment of evaporated water are recommended, especially if the tanks are not covered and illuminated by heat producing metal halide lamps. Of course, covering the aquaria will reduce evaporation but then we can not use halide lamps which renders the keeping of clams rather unattractive.

Ideal are devices which constantly monitor the salt concentration and automatically replenish evaporation losses by adding freshwater. Several models are on the market; some use electric pumps, others are operated mechanically with the help of gravitation. The latter have the advantage of operating without electricity.

Replenishment of lost water and preparing fresh seawater must be carried out with clean tap water. Pollution found in tap water will be tolerated by a number of invertebrates to a certain extent but giant clams in particular and molluscs in general react sensitively in the presence of heavy metals. Therefore, it is of utmost importance that the water used is free of such inorganic components like copper or iron. The local water suppliers can give the information wanted.

If a decontamination is necessary, the proper means is full de-ionization or reverse osmosis. These devices will produce water very similar to distilled water. Such water is ideal if we take into account that not only noxious substances but also necessary components are removed which must be replenished, especially calcium and hydrogen carbonate ions being indispensable for the formation of shells.

If decontaminated water is used for the solution of a good brand of sea salt, the resulting seawater will be very similar to the real thing but we should let the solution ripen for a few days as some components will not dissolve so quickly even if the optical impression seems to prove otherwise. During the process of solution toxic products will be formed which will harm the animals.

Only after a period of several days metal compounds will precipitate at the glass walls and intermediary products will reach their final composition. Now the water can be used without the danger of harming animals. The same holds true for the pH-value which will fluctuate during the solution process, but will then reach its final value of about 8.2 pH. Precipitated lime in the container used for preparation of the solutions must not be brought into the aquarium as it often contains toxic heavy metal compounds.

Not all reef inhabitants can be introduced to the reef tank, as demonstrated by these shrimps. They feed on a starfish Linckia laevigata, a scene that was seen between semi-adult clams in a clam nursery.

Lime reactor of KNOP-Aquarientechnik

The Calcium Concentration.

The highest amount of salt found in seawater is sodium chloride. All other substances are found in considerably lower concentrations and can therefore not be measured by relatively simple physical methods as it is the case with sodium chloride. The aquarium trade provides devices using chemical reactions, mostly of the titration technique, to measure these other components of seawater. The methods are normally very accurate, provided the expiring date of the kits is not surpassed.

If the supply with calcium and hydrogen carbonate ions is sufficient, the rocks will soon be covered with coralline algae.

153

Apart from sodium, an element of high importance when keeping giant clams in the aquarium is calcium. Many of the invertebrates we keep in our tanks possess either a shell or a skeletal frame made up of calcium carbonate. Even in many of the soft corals needle-like rudimentary skeletal structures are found, stabilizing the bodies and helping to detract predators. Contrary to snails or crabs which satisfy their need for calcium with the up-take of food, our corals and clams have to employ complicated chemical reactions to get the calcium out of the surrounding seawater.

To provide the animals not only with calcium but also with the necessary hydrogen carbonates, a relatively simple device is used which adds dissolved calcium carbonate into the tank water. The device is based on the method introduced by Hebbinghaus in 1994 (Hebbinghaus, R. 1994). The components of the original construction were altered in such a way to produce a compact and portable device. It does not need extensive maintenance and can be easily opened to be filled up with lime granulates. The reactor can be operated by introducing it into the circular system of the aquarium but it is also possible to run it with tap water purified by reverse osmosis. In this case evaporated water will be replenished by water with an appropriate contents of hydrogen carbonates. The principle of the system is rather simple; water is constantly circulated through calcium carbonate granules contained in the reactor and the solubility enhanced by reducing the pH-value. Due to the long reaction time the hydrogen carbonate contents of the water is high and a few liters per day suffice to add the necessary amount of calcium ions and hydrogen carbonate ions to the aquarium. The solid lime in the container will gradually diminish and the amount of carbondioxyd brought into the tank is so small that the alkaline buffer substances in the aquarium are enough to stabilize the pH. Nevertheless it is advisable to monitor the pH-value in the initial phase. The higher the carbonate hardness of the water, the greater the ability to bind free carbondioxyde and to stabilize the pH. To avoid an impairment of the milieu I advice to slowly rise the hydrogen carbonate and calcium contents of the seawater.

Natural seawater has a calcium concentration of about 420 milligrams per liter and a carbon hardness of about 7 degrees and we should attempt to reproduce this figures in our tanks. According to my experience calcium figures in between 400 and 450 mg/l and carbon hardness from 7 to 12 degrees are best suited for our purposes. Higher figures proved not to be harmful and are indeed possible when using the before mentioned instrument but do not bring any advantages.

The instrument shown in the picture can be filled with broken corals (calcium carbonate), or with coral sand. Rolf Hebbinghaus got the best results using coral sand with a gravel size of two to three mm. In my tanks I favour broken shells of the *Tridacnidae,* supposing that they contain what the animals actually need to build up their shells. More important than the material used, however, is a fine gravel size providing a large reactive surface between water and substrate. Small amounts of dissolved phosphate are not harmful as they will be metabolized by the symbiotic algae. In the case of higher and therefore harmful phosphate concentrations this can easily be remedied by introducing finely grained calcium carbonate into the biological filter.

154

A further way to enhance the calcium concentration in the seawater aquarium and being used worldwide is to add 'lime-water'. This was first used by Peter Wilkens in 1970. 'Lime-water' is a saturated solution of calcium ions and hydroxide ions. Such a solution is easily prepared; a tight container of about 10 liter (ideal are water jerry cans used for camping) is filled with distilled water. Now we add 100 to 200 gr calcium hydroxide, a white and somewhat clumped powder. After having tightly closed the container it is now thoroughly shaken to evenly distribute the powder. Left over night the superfluous powder will settled at the bottom and the water will be saturated with calcium hydroxide. With a plastic hose it can be taken out without disturbing the bottom layer. This solution can now be added to the aquarium, preferably drop by drop and near the jet of a water pump to avoid contact of the highly alkaline fluid with an animal.

The pH of the aquarium water should be monitored in short intervals when adding the concentrated calcium hydroxide. The solution can be added until the pH reaches the upper tolerable border. Further additions must be smaller and should be about two thirds of the initial dose in daily applications. Constant monitoring of the pH is of course necessary and inadvertent high doses can have a detrimental effect on the whole system. If adequately applied, however, these applications will stabilize the calcium contents of our tank and provide clams, corals and other invertebrates with the correct amount of calcium. In addition doses of calcium are replacing buffer substances neutralizing organic acids formed in the nitrification circle. Otherwise, bacteria would constantly produce acidic metabolites and thus eliminate alkaline substances causing a considerable drop in carbonate hardness and pH-value.

The remaining bottom layer in the container can be used four to six times to produce new lime water. After this the powder is exhausted and no longer able to produce lime water. We can recognize this state in the fact that no turbidity develops when mixing the clear lime water with our aquarium water. The bottom layer now has be replaced by new calcium hydroxide powder.

Calcium hydroxide of the highest purity must be used as cheaper preparations may contain heavy metals and other noxious substances and clams react especially sensitive to concentrations of these substances still tolerated by corals. The exact specific name of the chemical is "calcium hydricum purum pulvis". Also it is important to always tightly close the water container, otherwise carbon dioxide could transform our calcium hydroxide into calcium carbonate rendering our solution worthless.

Other methods to rise the calcium concentration in seawater have been proposed but the methods outlined above are the most practical, simple, and save. Other methods use different chemicals, for example calcium chloride with the release of chloride ions which can imbalance our system. We must never forget that our animals come from a very stable environment and their adaptability is not capable to cope with, let us say, sudden rises in pH from 8.0 to 8.6 or more. Even the smallest doses of chemicals introduced into our aquarium system are huge in comparison to the stable marine environment and the animals' regulatory systems are not adapted to such sudden changes. We can look at our own body with its 100,000 chemical reactions that take place simultaneously. Every medicament will induce innumerable changes in regulatory systems influencing each other. A normally harmless drug can sometimes induce reaction chains similar to a domino play with the possibility of a high number of disadvantageous side effects.

Similarly our aquarium is a highly complicated system of inorganic and organic chemical reactions. An explanatory attempt will be a rough simplification of the real state of affairs. Clams are highly sensitive to environmental changes and any experiments should be avoided.

Carbonate Hardness and pH-Value

Apart from the calcium contents and the pH-value of the aquarium water, the acid binding ability must be regularly monitored. The binding of acid uses up carbonate as well as hydrogen carbonate and these constituents must be replaced in regular intervals. If we use the calcium reactor described above, this will be accomplished automatically as the reactor will replace the right amount of hydrogen carbonate ions and we will not have any difficulties in keeping the pH-value constant. The pH-value can easily be checked with the help of kits which are available in the aquarium shops. The pH-value should be in the range of 8.0 to 8.1 but values of 7.9 and 8.2 are the lower and upper limits. It is true that most of the invertebrates we keep in the aquarium will tolerate values up to pH 8.5 or even more but giant clams will not because they will show signs of discomfort if confronted with pH-values under 7.9 or over 8.2 (Huguenin and Colt 1989). Here we encounter one of the problems in keeping clams.

In case of a low pH-value which cannot be rectified with the calcium reactor or by adding lime water we must think of a source of high acid production in our tank. Higher amounts of debris hidden underneath decorative stones is one of the more frequent causes for low pH-values. Such debris collecting crevices must be carefully cleaned. A part exchange of the aquarium water is also recommended.

In case of a high pH-value, however, the addition of calcium in any form is contraindicated. The apparatus described above will rectify this condition and, if constantly in use, will guarantee the optimal pH-value.

The acid binding ability of our aquarium water which is governed by the carbonate and hydrogen carbonate contents as already mentioned can also be measured using a chemical test. The carbonate hardness test tells us the amount of carbonates present expressed in 'Grad deutscher Härte' (dKH). The hydroxide ions present in our lime water will also be recognized as carbonates in this test. Natural seawater has a dKH value of around 7. If we find a value under 7 in our tank, appropriate means must be taken to bring the dKH value up to 7. If we find a value greater than 7 this will not be harmful as the production of acid matter in our aquarium is much higher than in the reef. This is the reason why it can be advantageous to keep the dKH value up. 12 dKH should be enough but higher values up to 20 degrees will be tolerated without any adverse effects. However, low values of 1 to 2 degrees dKH will certainly pose a threat. The ability to bind acid will be so much reduced that a sudden drop in the pH-value cannot be buffered and a sudden drop will follow resulting in an imbalance of our milieu in the tank. The buffer capability must then be improved.

Trace Elements

Clams, stony corals and coralline algae need further substances to turn dissolved calcium into a solid state. Apart from strontium, molybdenum is needed to turn calcium carbonate into the crystalline aragonite. Organically bound iodine is another trace element, indispensable in the aquarium where it quickly can reach low levels. This elements can be added to the aquarium water individually or in form of a concentrated mixture. The advantage of adding the individual trace elements is that the needs of the clams can specifically be satisfied as they are different from the needs of soft corals. After all we know about invertebrates and fishes of the reef, the so-called 'macro elements' like sodium or potassium are as important as the 'micro elements', which are the trace elements. Very little is known about the action of most of the trace elements but they are certainly indispensable for many of the metabolic pathways of our animals in the tank, even if they can be highly poisonous in higher concentrations like copper and others. Therefore, the trace elements lost through the action of carbon filters or skimmers or even natural losses must be regularly replaced. My own experiences with CombiSan, developed by Peter Wilkens at the end of the 1970s, are very good. CombiSan, a concentrate of trace elements, also contains organically bound iodine, strontium and molybdenum.

In case of vitamins we find similar conditions. Very little is known about their importance for fishes and invertebrates of the reef. However, practise has shown that regular doses of vitamins have a beneficial effect on the colouration and reproductive rate of our animals. As I said before, very little is known about vitamin action in reef animals but the fact that regular doses of concentrated vitamins can inhibit the fading of colours in coral reef fish like *Paracanthurus hepatus,* contributed by some authors to advanced age, is proving their importance. The same holds true for the fading of colour in the purple *Pseudochromis purpureus* which can effectively be prevented by regular doses of concentrated vitamins. But not only discolorations can be prevented by the action of vitamins. Degenerative alterations of the skin leading to scar formation in older reef fishes can be kept at bay with regular doses of vitamins. Such degenerative alterations might be caused by the action of parasites like *Hexamita/Spironucleus* infections and it seems likely that vitamins play an important role in enhancing the immune system. Older reef fishes of the genus *Acanthurus* seem to prove this hypothesis. They need a high percentage of agal food to satisfy their need on vitamins which cannot always be satisfied under aquarium conditions. Discoloration and a high susceptibility to infectious diseases will follow.

Clams need higher concentrations of vitamins than fishes. The latter take vitamins up with the food but the former must take them directly from the surrounding water requiring much higher concentrations. The concentrate manufactured by Kent under the trade name 'Zoe' proved to be especially effective but there are several other products on the market. The products are always accompanied by a detailed instruction how to use them.

Water Temperature

As all other animals from tropical regions, our clams need heating the aquarium water if kept in higher latitudes. The *Tridacnidae* react sensitively to sudden changes in temperature and the optimal range is within 25 to 32 degrees Celsius (Huguenin and Colt 1989). 22 and 34 degrees Celsius are the lower and upper limits and should be avoided.

The danger of loosing costly animals due to a sudden drop in temperature or a faulty thermostat has led me to a sort of security device by using two independent thermostats and heaters where the second heater will automatically come into action if the water temperature drops below 23 degrees Celsius. Such an arrangement will prevent cooling with consecutive loss of animals, provided, of course, the electricity does not fail.

In summer, however, even at our latitudes, a considerable rise in temperature can occur and counteractions become necessary if the temperature rises above 30 degrees Celsius. Heat producing devices like pumps must then be switched off. Heating produced by the metal halide bulbs can be reduced by shortening the illumination period or by enlarging the distance between bulbs and water surface. If these measures do not help, evaporation with the consecutive drop in temperature can be achieved with the help of an electric fan. If all this will not produce the desired effect, ice cubes enclosed in a plastic bag and brought into the tank will certainly change the situation. During these manipulations a sufficient saturation of the aquarium water with oxygen must be maintained. Water with a high salt concentration will not contain as much dissolved oxygen as water with a lower salt contents and lowering the salt concentration to the tolerable value of 1.020 will help to master the situation. The success of such a measure can be seen in the less heavy opercular movements of the fishes. Of course, equipment of a high accuracy is necessary if we are going near the lower level of salt concentration. It is important that the protein skimmer is in good order as it helps to eliminate undesirable gaseous components and rises the level of oxygen. If the tank does not contain a skimmer, aerating the water through a diffuser block is essential.

Giant Clams in Reef Tanks

Keeping clams in a community tank normally does not pose any problems. They do not grow in an invasive way by dislodging other invertebrates and they do not possess poisonous substances. On the other hand they seem to be rather insensitive to the touch of corals. Even direct contact between the siphonal lobe and the irritative tentacles of corals like *Zoantharia* as well as *Octocorallia* does not seem to cause any discomfort in the clams. The only problem can be caused by excessive growth of soft corals with the subsequent formation of shadowed areas. Mighty individuals of *Sarcophyton,* growing to a size of over 60 cm diameter, species of soft corals like *Cladialla* and *Lithophyton* with profuse growth are able to harm clams or groups of them by holding back the light. In such cases I usually move the soft corals to different areas rather than moving the clams as the latter often react more sensitive to changed light conditions than the corals. I have the impression that the *Tridacnidae,* as soon as they have reached maturity, are much less adaptable to changed light conditions than soft corals from the shallower reef areas. Therefore I leave the first choice of location to the clams and allow compromises only with soft corals. If a soft coral shows signs of discomfort, a change of location in the tank will often bring back the former state of good health, even after a considerable length of time. In the case of clams, however, such a manoeuvre can cause sudden death of the animal. The cause is probably the clam's effort to make up the lack of light by rising the number of symbiotic algae. To facilitate the growth of the algae the clam will open the siphonal mantle to the utmost extent, giving the uninitiated observer an impression of great health. On the other hand, moving the clam to a brightly illuminated area can also cause the animals demise in a few days time. Braley (1992) described the complete shading of a *Tridacna gigas* of 20 cm length over a period of five weeks only to die suddenly when brought back to the former full exposure to natural sunlight.

Keeping clams together with fishes, however, can prove difficult as many coral reef fishes will more or less constantly molest the clams and can cause considerable harm. This does not necessarily mean that the fishes feed on the clams. The cleaner fish *Labroides dimidiatus,* following his natural instincts, will constantly nibble at the siphonal mantle, picking at colour patterns resembling parasites and thus constantly molesting the animal. In the chapter about clam diseases these phenomena were described in more detail. The fishes' behaviour not only causes injuries in the clams' mantle but constant irritation and the clams will only partly open the mantles. On the other hand, the fish will not learn from it's error as it follows it's natural instinct. Sooner or later, the clam will succumb to the constant irritation if the fish is not removed.

Similar effects can be caused by *Chelmon rostratus,* though this is not a cleaner fish. Often the clams' iridophores are attacked by this species and though not all individuals of *C. rostratus* will behave in this manner, no predictions can be made. Often one individual will live in perfect harmony with the clams only to change suddenly it's behaviour and will vigorously start attacking them. This is why I do not recommend keeping such fishes in the presence of giant clams. Of course known predators of mussels like *Labridae, Chaetodontidae,*

Pomacantidae, or *Balistidae* must not be kept together with clams. Even obviously peaceful specimens of emperor fishes mostly start to attack the mussels after a certain period of time. Often it is very difficult, even with the aid of fish traps, to remove such individuals from the aquarium and many small organisms flourishing in the tank provide enough food and making a baited trap not a tempting proposition.

It should be pointed out that certain labrids could play a beneficial role in the invertebrate aquarium. Years ago I recognized that juveniles of *Coris gaimard* can be used to fight parasitic slugs. Aquaria heavily infested with such parasites could be freed in a short period of time, because one young *C. gaimard* only was enough to considerably reduce the juvenile parasites while the adult ones, being too big for the fish, could be easily removed by hand. Though it was not possible to wipe out the parasites completely as there was always a small remaining population hidden under the rocks and corals, but their number was too small to cause any damage in the corals. Only after the demise of the *C. gaimard,* the parasitic population started to grow again resulting in considerable damage to the soft coral colonies in the tank. I know of four such cases and as *C. gaimard* also feeds on snails, it should be possible to control carnivorous parasitic snails harmful to clams. In the chapter about clam diseases this problem is dealt with in greater detail. Adult individuals, however, should not be kept in the invertebrate aquarium. *C. gaimard* is a rather rugged fellow with conspicuous dentition when grown up and should not be trusted.

There are, however, other foes for the clams in our aquarium. It is not uncommon to find bristleworms inside dead clams feeding on the remains and their detrimental influence is well known among aquarists. Therefore clams should not be kept on sandy bottom but over a rocky substratum as the former facilitates the entrance of the worms through the byssus apperture. I myself doubt the efficiency of this advice because the bristleworms ability to negotiate narrow clefts is remarkable and hardly any clam will be sealed completely around the byssus aperture, even on a rocky surface and there will always be some small clefts around the aperture where the worms could enter. There is probably a biochemical defence mechanism which aggravates the entrance of the worms into the clam by means of the secretion of specialised glands situated around the byssus apperture but this is only a supposition without scientific verification. In addition it is not completely clear if the worms are really parasites or if they enter the clams after their demise. This is still an unsolved question. Bristleworms are highly sensitive to smell and are probably attracted to dead clams by the smell of decaying tissue. Clam tissue disintegrates much more quickly than a dead fish. It could also be that an outwardly healthy looking clam contains already some decaying parts as the process does not affect the animals tissues simultaneously. The bristleworms would then enter a clam already destined to die but with an outwardly still healthy appearance. There are hints to such occurrences and shrimps and fishes will start to behave conspicuously. For instance, the shrimp *Hippolysmata grabhami* will suddenly tear off junks from the siphonal mantle of a clam with still healthy appearance. In reality, however, small decaying parts of the diseased clam will attract the shrimps, fishes, and other organisms feeding on carrion.

Surgeon fishes, like this pair of Zebrasoma flavescens, are well suited to the reef tank. The clam in the centre of the picture is Tridacna squamosa.   *Photo: Klaus Jansen*

A moray eel with its mandible securely locked by the shell of a Tridacna crocea. Mishaps like this are mainly due to carelessness of the fish.
Photo: R. Wong

A juvenile of Coris gaimard. It can be a useful inhabitant of the invertebrate aquarium.
Photo: H. Göthel

Tridacna gigas displaying the mantle with its characteristic colour pattern.

Other Zebrasoma species like Z. xanthurus or Z. gemmatum can also be kept together with the Tridacnidae.
Photo: Klaus Jansen

An other situation is probably caused by excessive multiplication of a bristleworm population. Shear lack of food would probably cause the worms to attack healthy clams and acting as pace makers for other organisms like shrimps and fishes. In such a case it would make little difference if the clam stands on sandy bottom or on rocky substrate.

Feeding Clams

It is a widespread misconception that clams live solely on their symbiotic algae. The presence of a fully developed intestinal tract with digestive glands, bowels, and anal papilla as described in the chapter on anatomy speaks for the contrary. As all molluscs they are dependent on food dispersed as fine particles in the water which are then filtered through the gills. They are very particular to the size of these particles as bigger pieces will be selected and expelled as clumps. By a contraction of the shell reminiscent of a cough these clumps will be discarded. Only the smallest particles present in the plankton, the so-called nanoplankton, is accepted as actual food. Not only the size of the particles but their concentration in the seawater is also important and concentrations too high will provoke the same reaction as already described for particles too big. This is why it is not possible to feed clams with the help of a pipette. The whole metabolic mechanism of the clams is not designed to cope with sudden large amounts of food. This, by the way, holds true for most of the reef animals feeding on filtered food. Feeding clams should simulate the natural condition with small concentrations of food available over lang periods of time.

Therefore, clams should be feed daily or at least every second day. If kept in an aquarium populated with a large number of other invertebrates, feeding on finely suspended food, the clams do not need special treatment. If so-called 'floating food' is used, however, clams need additional feeding as the particles will be to large for them to be filtered. It is better to feed the *Tridacnidae* on food based on yeast. Yeast consists of spherical or ellipsoid fungi with a fast proliferation rate and a high contents of vitamins and protein. They make an ideal food source for the clams being very similar in size to the natural food supply. Life yeast as used for baking is best though rather perishable. Even better is suspended yeast especially prepared for our purpose. Yeast must be sparsely used. For my own tank holding 6000 liter seawater and being populated with a large number of invertebrates like stony and soft corals and numerous clams, yeast in the amount of a hazel nut is sufficient. It is not so easy to get the right amount of food for let us say a tank of 250 liter. It is best to prepare a small amount of yeast, of which only a small portion is put into the tank discarding the rest. Altogether it seems easier to use commercial preparations which in addition may contain other important ingredients.

I have tried several suspensions made up of other raw materials, some of them more or less successful, but the yeast based suspensions still seem to be best. One of the more successful experiments was carried out with blood. Blood cells are similar in size to the nanoplankton. In addition blood contains a vast amount of nutrient and due to the salt concentration is very similar to the osmotic conditions found in seawater leaving the individual blood cells intact when brought into the tank. The application is similar to yeast and a few drops suffice to cloud a large aquarium.

The reaction of the different groups of invertebrates was very positive showing prolific growth rates and expansion. However, the high surface tension of the blood cells caused a high rate of skimming which was counteracted by switching off the protein skimmer for about one hour after feeding. Only a short while after the reactivation of the skimmer, the water had cleared and the pots collecting the foam where full. Easy and quick removal of superfluous amounts of food by skimming seems to be a further advantage of this sort of food supply.

I stopped these experiments because of the danger of various infections cannot be completely excluded, even when using animal blood. An other negative point is the short conservation time and any attempt to store blood would result in a considerable drop in the contents of nutrients.

In the absence of the possibility to carefully conserve and sterilise the blood intended for use in our aquarium as food, I do not recommend this method. Probably the aquarium industry can develop food along these lines in the future. Sometimes it proves advantageous to try unconventional ways. I know of a pathologist fertilizing his garden plants with expired blood conserves to great success.

If using yeast as food for the invertebrates, the governing factor is the size of the tank and not the number of animals present. We have to keep in mind that only a small percentage in the range of one to two per cent of the suspension is actually used as food and the rest eliminated by the filtration system, most of it by the skimmer. Such a degree of 'over feeding' is unavoidable to obtain the food density necessary to satisfy the food requirement of all the animals. This "over feeding" should be kept as small as possible. Corals and clams are adapted to a slow food up-take and therefore a daily but small amount of food seems to be better than weekly and larger doses.

Clams do not only take up the food we introduce to this purpose but utilize other matter as well as, for example, waste products of fishes present in the tank. Altogether the food requirements of clams is so small that a special feeding scheme for them is superfluous if the tank is populated with a fair number of fish, especially if corals are regularly fed. In such a case a special food destined for the clams could even unbalance the whole system.

Another situation, however, is caused by the symbiotic algae living in the siphonal mantle of the clams. Their population density per square unit is about ten fold in comparison to corals producing a high degree of photosynthesis and it follows that the need for organic nitrogen as fertilizer is equally high. If only a small number of clams is kept, the amount of nitrogen utilized is probably negligible but larger numbers of clams as found in the tanks of some aquarists or dealers will certainly lead to nitrogen deficiency with consequential impact on the well-being of the symbiotic algae. The symptoms have already been described in the chapter on clam diseases. Under such conditions it is necessary that we provide the algae with the nitrogen needed. The Philippine biologist Carmen A. Belda investigated the influence of nitrogen fertilization on the growth rate of giant clams in an excellent paper published in 1993 (C.A. Belda 1993 b). After adding nitrogen to the tank water a significant raise in the growth rate of the clams' soft tissue had been observed, the number of symbiotic algae rose drastically with a concomitant fall in the chlorophyll -a contents of the individual zooxanthellae. This result is in concordance with other investigations about the phytoplankton in the presence of higher levels of nutrients.

From this it should be clear that the old notion of the noxious effect of nitrogen and phosphate in the aquarium water must be seen in a new light. If our clams show signs of discomfort which could be traced back to a lack of this nutritiment as described in the chapter about clam diseases, we must artificially add them to the aquarium water. For this we use a 0.1 % standard solution of sodium nitrate (analytical quality, Merck). One gram sodium nitrate is dissolved in 1000 ml distilled water. from this solution we add not more than 10 ml per 100 l aquarium water. The nitrate contents of the water must be constantly monitored with the help of a precision instrument and must not raise over 2 mg/l. The importance of these measurements should not be neglected. An overdose of the standard solution could seriously imbalance the milieu of our tanks. Our aim is to eliminate a lack of nitrate but not to raise the nitrogen level to an unphysiological level. The success of such a measure can be seen in the deep brown colouration of the mantle growing from outer to inner areas.

Carmen Belda (1993 B) further investigated the effect of higher levels of nitrogen on the growth of the clams' shell. The growth rate is drastically enhanced but at the same time structure is altered and becomes porous and nearly transparent with a much lighter overall weight when compared with a normal shell. Electron microscopic investigations reveal changes in the crystalline structure of aragonite and a porous architecture.

As Rasmussen (1989) has shown, very similar alterations can be observed in stony corals if kept under similar conditions. The corals used for this study were *Acropora formosa* and several species of *Porites*. If kept in tanks with a high contents of phosphate, the corals' skeleton becomes porous with a weak structure of aragonite. These results were confirmed by further studies (Kinsey and Davies 1979, Kinsey 1991).

From the before mentioned facts it should be clear that the nitrate concentration must not surpass certain levels.

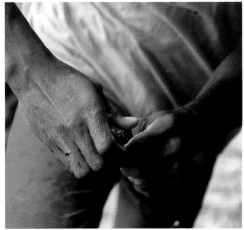

Tridacnid clams are often eaten as a „fast food"
by the fishermen. They sometimes open it with
their harpoon direct after catching and eat it
raw.

Selection of Suitable Sites and
How to Detach Byssus Threads.

As already indicated, the clams' ability to adapt to an altered environment is not unlimited especially if we are dealing with animals which have been only recently brought into the aquarium. Originating from the low water areas of a hatchery, they are adapted to a special environment. Their juvenile phase with its greater adaptability is over and they should have found by now a site of constant and stable conditions. Their physiology is not designed to cope with drastic changes of the environment. It is true that during the rainy season the clams receive less light because of the cloudy conditions but the change from sunlight to artificial light with its different intensity and spectral composition especially after a dark period of 30 to 40 hours darkness during transport certainly stresses the clams' adaptability to the utmost. The condition in the dealer's tank with mostly unsatisfactory light conditions only adds to the hazards. Nearly all the parameters the animals are adapted to will be altered; gaseous metabolism, pH-values, and many others. It does not take wonder that some individuals will react in a confused way, rejecting the new conditions altogether. Often it is difficult to make out the negative factors as the site rejected was probably hold before by another clam showing not signs of discomfort at all. If the new site is not accepted by the clam by dislocating itself with the help of vigorous movements of the shells, the reason may well be extreme differences in environmental factors as compared to the ones experienced in its former life.

On the other hand it could be that the clam is perfectly aware of the adverse factors. Often the reason is a change in the quality of light, either in intensity or spectral composition. If the light intensity is low, no adverse reactions have been observed. In case of high intensity, however, signs of discomfort are not uncommon. The grade of light intensity is seen here in comparison with the conditions prevailing before the clam's captivity. After removal of the animal from the hatchery during the rainy season with it's overcast skies and having stayed in darkness for at least 30 hours during transport, a new location under a strong metal halide lamp will certainly strain the clam's adaptability. It will then try to hide it's mantle in shadowed areas under corals or will escape from the chosen location altogether by vigorously closing it's shells. In such cases we must find out the adverse factors to be able to alter them. If new clams are to be introduced into our tank, they should never be placed directly under the metal halide lamp. If the animal is intended to be placed at it's final location the area should be shadowed by means of a frosted glass plate for several weeks. In an emergency normal glass with sprayed-on salt water will suffice; the salt crystals will then disperse the light.

If the intended location is not accepted, other factors than light conditions must be considered. Sometimes the animals experience difficulties in fixing their byssus filaments to the substrate. The reason could b the substrate's surface. Sometimes it helps to move the animal a few centimeter to a different location with a seemingly better structured surface. Water quality can also play an important role and a ph-value too high can trigger the before mentioned behaviour. If salt concentration, ph-value and other parameters are in their respective tolerances the clam will accept its new location in a few days or weeks time. It should not be moved from the chosen site. If it has fallen from the decorative rocks to the bottom of the tank and seems to feel well with open mantles we should leave it under the precondition that the illumination is satisfactory. The physiological requirements of the clams must prevail over esthetic considerations by the aquarist.

A frequent problem is the removal of a small piece of coral on which the byssus filaments are fixed as it will cause the clam to turn to its side if located at its intended site. The hatcheries often dispatch the animals with these small coral stones attached to demonstrate healthy byssus filaments. In the chapter on clam anatomy I have pointed out the delicate structure of the byssus gland.

If it is not possible to find a suitable site due to the coral fragment attached to the byssus filaments it is advisable to carefully remove it. It is not true that the removal of such a coral fragment leads automatically to the death of the clam. I have carried out this procedure on more than 50 individuals without one casualty. A short and clean kitchen knife is carefully pushed between the stone and the shell and the filaments successively cut. To do this the stone is only slightly pushed to one side to free it from the byssus apperture until the first filaments become visible. They must be cut close to the stone not to injure the byssal tissue. If the clam tries to protect the filaments by expanding soft tissue, a slight touch with the knif's back will cause it to retract the glassy whitish tissue. This procedure can be carried out outside the tank without causing any harm to the clam. This is better than doing it inside the tank necessitating acrobatics on the side of the aquarist with the danger of harming the clam.

Sometimes the small stone is so tightly fitted into the byssus apperture that its removal is impossible. Supporting the clam with additional flat rocks is a better way to prevent them to fall to the side.

Sometimes it is necessary to cut the byssus filaments of clams living already for a long time in an aquarium. Fast growing species like *T. maxima* and *T. squamosa* will eventually shade each other by the growing mantles causing total bleaching in the shaded areas and necessitating a relocation. Especially small species of the *Tridacnidae* tend to tightly adhere themselves to the substrate and removing the shells cannot be accomplished without severing the byssus filaments. Such an operation should be carefully considered and if unavoidable, very carefully carried out in the way already described. The animal should be slightly tilted till the byssus apperture is lifted from the substrate. What we see first is the soft and swollen body tissue. If slightly touched with the back of the knife, it will be retracted and the byssus filaments become visible. They can be then carefully cut. The more of them are separated, the less pressure is necessary to tilt the clam.

A dense and varied population of invertebrates in a large aquarium.   Photo: Enrico Enzmann

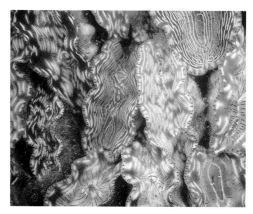

Specimens of one of the most colourful clams, Tridacna derasa, bred at the UPMSI, Philippines.

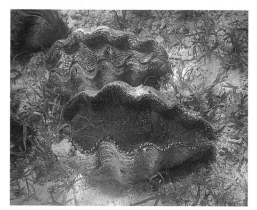

Two nicely patterned Tridacna derasa, bred in a clam farm.

Tridacna crocea and T. maxima on artificial substrate.

A display tank can be an attractive eye catcher in business or clinic rooms.   *Photo: Klaus Jansen*

Buying Clams in the Pet-Shop

Good health is a prerequisite for well-being and longevity of our clams in the aquarium. When buying clams certain precautions must be taken. Are they kept under appropriate conditions? Is the illumination sufficient? If all this is not the case, the purchase is not recommended.

The clams must give a healthy impression, the mantle must be fully opened with lively colouration throughout. Areas of transparency, especially in the central areas of the mantle should raise suspicion. The same holds true for the shells of dead clams amidst live ones in the dealer's tanks. As a matter of fact, I do not recommend to buy freshly imported animals which have not recovered from the stress caused by the long journey. The animals should have had time to adapt themselves to the changed conditions before being again confronted with a new change in milieu.

A rocky substrate is not a prerequisite for the well-being of the clam. The animals can be placed in sand with their byssus filaments without impairment of their health. However, if we can find animals attached to small stones, this should be considered a sigh of good health and the small stone can be a great help in the decision to buy.

Shells must be thoroughly inspected. Broken scales are of little importance and rather a flaw in beauty than a serious matter. A search for parasites like parasitic snails, fungi, or mussels as described in the chapter on clam diseases, however, is important. An other important test is the clams reaction to shadows; a swift retraction of the mantle when a shadow is produced by moving ones hand over the shell is a further sign of good health. If the reaction is weak or lacking altogether, the animal is certainly in a poor condition.

Dealers should illuminate the tanks holding clams with halide lamps but not with fluorescent tubes. Clams can be kept either in community tanks together with other invertebrates or separate aquaria. In any case must the seawater be absolutely free of heavy metals or residues of other chemicals like medicines. I know of a case of heavy metal poisoning in which the residues of copper from a treatment of fishes in the past caused numerous casualties in clams but was harmless to the other invertebrates present in the tank stressing the vulnerability of the *Tridacnidae* in comparison to other organisms. An other important factor is the pH-value which must lie in the physiological range. Damage due to a highly alkaline water can normally not be rectified in clams. In this respect, the *Tridacnidae* are much more delicate organisms than corals from shallow reefs with their much higher ability to regenerate. Keeping clams needs a higher degree of judgement than keeping the before mentioned corals.

Closed Aquarium Systems

A closed aquarium system is a system without connection to the open sea; it functions as an isolated unit. As already mentioned in the chapter about keeping clams in aquaria, such a system posed problems not present in the reef. The amount of metabolic turn-over in relation to the water volume is very high. The organic waste produced in a closed aquarium system is much too high to be regulated by natural pathways and we need technical devices to keep our miniature reef in a stable equilibrium.

Most of the closed system aquaria are display-tanks, designed and decorated along esthetic lines. This is perfectly alright with the form and construction of the tank and the decorative rock structures. This does rarely pose problems as long as the requirements of the animals to be kept are met and certain physical features are present. The water surface of the tank for example should be larger than the front glass. The height of the water column must not be too high so that enough light can penetrate to the bottom of the tank. This can be remedied with the use of metal halide lamps in higher latitudes but the great heat production of these lamps poses problems in tropical countries. In Asia I have seen tanks which had to be illuminated with fluorescent tubes but not halide lamps during the hot season to prevent overheating. A great height of the tank would be disadvantageous under such conditions.

Other considerations become important if we want to decide which animals we are going to keep in our tank. If we want to keep fish we must know which species are compatible and what the requirements are. The same holds true for invertebrates, no matter if we think of shrimps, crabs, or sea urchins or sessile corals and anemones.

It is customary to populate display-aquaria with a variety of organisms. As many of the animals we keep occupy different ecological niches with different behaviour pattern, a rather large diversity of species can be kept without molesting each other. Such a diversity makes good sense and will reduce competition for food. The fertility of the individual species will not be as prolific as when only one coral species is kept in a tank, in as much as many corals produce certain secretions which inhibit the growth of others. However, the differing requirements of the species will help to keep up a balanced state and will prevent to produce a bias in certain components in the seawater. And last not least, a reef tank is much more attractive to the observer than the monoculture of one species of coral.

The same holds true when keeping clams. I have experienced serious problems when keeping solely *Tridacnidae* in the aquaria. The competition for organic and inorganic nutrients in clams is so fierce that deficiencies will rapidly develop in a closed system. The animals will stop to grow and become susceptive to a whole range of health hazards and infectious diseases are difficult to control under such conditions. The best conditions for keeping clams are in a reef tank with a broad variety of other invertebrates as well as fishes. The population density of clams should not be high, especially if we are keeping large specimens. In a small tank one large clam is able to imbalance the system by extracting too many nutrients and trace elements if these are not constantly remedied by changing the seawater or replenishing the trace elements.

Construction of an experimental tank by KNOP-Aquarientechnik.

Decorative rockwork fixed with concrete (iron-free Portland Cement, watered for two weeks).

The same tank as above after one year. Several corals had to be removed during this period because of fast growth.

Large aquarium of 14.000 l. The decoration is not yet completed. Photo: K. Jansen

The technic of the aquarium shown above.

Large Aquaria

If the aquarium exceeds a certain size, special problems will arise. One of them is the biological inertia of the system. The advantage of this is that changes in the milieu will be slowed down in contrast to the situation in a small tank. For example, if an animal dies the sudden increase in waste products will more easily be dealt with by the biological system in a large tank than in a small one. If, however, we want to get rid of an overgrowth of lower algae in a large tank by altering the milieu, elaborate manipulations become necessary. Often such attempts involve considerable physical strain. To clean a small tank from filamentous algae with a brush is child's play but an aquarium of five to ten cubic meter will pose different problems. Mostly it is not enough to roll up one's sleeves to carry out certain tasks but one has to actually step into the tank.

This is why the necessary maintenance and manipulation of the biological system are often neglected with the result of an imbalanced milieu. Therefore, the biological as well as esthetic value of such show aquaria are often below the smaller "living room tank" looked after with loving care by their owners. The recent success and development in the keeping of marine organism have all been gained through carefully maintaining and observing such "living room aquaria". Certain tanks in public institutions give a lively picture of the difficulties encountered when maintaining such huge aquaria. The beauty of an aquarium is not necessarily proportional to it's size. To the contrary; the smaller the tank, the more attention is given to details.

An other disadvantage of large tanks is their immense financial cost. A regular change of seawater or the replenishment of trace elements turns out to be a costly exercise. But if expenses are cut, the biological equilibrium will not be maintained, animals will suffer from malnutrition and other deficiencies, and the esthetic as well as biological value will drop. The same holds true for the light sources and if not satisfactory, stony and soft corals will never flourish. Especially private individuals easily run the risk of stressing their financial possibilities to the limit.

If aesthetically decorated and kept in a biologically stable milieu, however, an oversized aquarium can be of striking appearance and a glance through the front glass of a tank of two to three meter length can make an unforgettable impression. Of course, clams as well as the other invertebrates will not care about tank size but the overall impression of such an artificial reef with it's numerous species of invertebrates and fishes displaying natural territorial behaviour can be breathtaking. The smaller species of the *Tridacnidae* will not be as conspicuous in a large tank as in a small one but the marvellous reef landscape will compensate for that.

The Tridacna Aquarium

Some aquarists want to keep aquaria especially designed for the *Tridacnidae*. Illumination, filtration and fish population are all designed to suffice solely the requirements of clam. One can argue about the aesthetic value of such a venture, favoured by many aquarists.

Such a monoculture will certainly pose certain problems. Parasites specific to clams can spread quickly under such conditions. Pyramellids, small snails parasitizing on clams at night, can be brought into the tank as a batch of eggs well hidden on the shell of a newly introduced individual. Infectious diseases like the white spot disease (WSD) or bacterial infections causing necrosis of tissue will spread in no time if inadvertently introduced. Decontamination of the seawater using an ultraviolet lamp over several weeks is a powerful tool to fight such hazards.

Another difficulty encountered in such specialised aquaria is the constant removal of trace and main elements. All clams are in need of the same substances leading to fierce competition. Adult individuals are better equipped to cope with the situation than juveniles which still need relatively large amounts of certain substances because of their intensive growth rate. If calcium, hydrogen carbonate and other elements necessary for the formation of the shell are lacking, the clams will stop growing. The whitish fringes of the shells as signs for new apposition of chalk cannot be longer observed. This is an alarming sign and must be immediately remedied either by changing at least 10 % of the seawater in weekly intervals or by use of a calcium reactor and appropriate doses of trace elements like CombiSan.

Food deficiency is another hazard encountered in such aquaria. The few fishes normally introduced will not produce enough organic waste to satisfy the metabolic requirements of the clams with their high food uptake in comparison to corals. A low concentration of nitrate and phosphate compounds is the result if not constantly replenished by nitrate contaminated tap water used for dissolving the sea salt when partchanging the water. I have described in detail the consequences and symptoms of food deficiencies in the chapter on clam diseases.

Despite the hazards mentioned, an aquarium solely destined to hold giant clams is still an alluring enterprise. I have constructed an aquarium which takes into account the fact that clams normally look much more colourful when looked at from above. In this tank the upper half of the front glass is bevelled to 45 degrees allowing one to view nearly the whole contents from above. Such a construction does not only bring out the beauty of the clams to advantage but that of fishes and corals as well. The diminished surface with a smaller area for gaseous turnover is compensated for buy an especially powerful skimmer. Many of our customers are very enthusiastic about this design and it may be that ingenious aquarists will come up with similar constructions in the future.

▲ Fossilized Tridacna squamosa, fully stoned with shell lenght of 30 cm.

▼ Photographing tridacnid clams works best with camera and flash coming from the same direction, usually from above.

The same aquarium viewed through the lower, vertical front glass...

... and through the upper, sloping half of the plate.

Purposebuilt Aquaria

Apart from aquaria constructed and decorated along esthetic lines, others are constructed to a special purpose. Form, size, and contents must follow here other criteria. As an example I want to describe aquaria I have constructed several years ago for the sole purpose of propagating corals. It runs on very little electricity. A pump in the lowermost tank brings water to the upper aquaria. The surface water of each individual tank then flows back through a single tube connecting the tanks as shown in the picture. Biological filtration is also located in the lowermost aquarium. Every one of the very shallow tanks is illuminated by one or two fluorescent tubes (daylight type, ballast externally located to reduce heat). It is wholly

Shallow, single tanks used for culturing corals.

constructed from glass plates eliminating the use of a rack and keeping evaporation to a minimum. Ten mm glass is used for the lower tanks and eight mm glass for the upper ones. Using different dimensions, this construction would make a good arrangement for a pet shop. It would be of limited use, however, if the dealer wants to display clams. The low light intensity obtained by fluorescent tubes excludes the keeping of clams, especially the smaller species like *T. crocea* and metal halide lamps cannot be employed due to the lack of space above the tanks. A further disadvantage is that the tanks cannot be viewed from above. The ideal tank to keep clams at a dealer's shop must be constructed along different lines. It should be wide and deep. 25 to 35 cm water depth and an area of one to two square meters would be appropriate to hold a large number of clams together with their stony substrate. The illumination could be cost effectively carried out with one 250 W halide lamp per square meter. Stronger lamps are not recommended because of the low water level.

To avoid deficiencies in the water quality due to the high density of clams with their specialized requirements, the circulation of such a tank should be connected with aquaria holding corals. The sperm production of the clams alone could seriously imbalance such a system with only about 300 liter per square meter. The lack of oxygen would soon lead to the death of some of the clams with further worsening of the water condition and a catastrophe would follow soon. The connection to a larger aquarium populated with corals would enlarge the overall volume of water, provide a certain amount of organic nitrogen to feed the symbiotic algae and prevent excessive consumption of trace elements. In case

of a similar volume of the two containers an hourly turnover of the tank water would eliminate the hazards described above, provided the filter systems are adequately dimensioned.

Tanks holding clams intended to be sold should not contain fish with the probable exeption of one juvenile individual of *Coris gaimard,* provided the mussels are not too small not to rise the interest of the labrid. *C. gaimard* is very useful to control parasitic snails of the Pyramidelline family. The snails are only active at night and could cause serious damage in a tank solely populated by giant clams if not effectfully controlled.

If trace elements, calcium and hydrogen carbonate are regularly replenished and if care is taken not to let the nitrate concentration sink below 2 mg/l, such a tank will keep giant clams healthy over a longer period of time.

Keeping Clams for Consumption in Restaurants

A completely different type of tank is needed when keeping giant clams for consumption. I have designed such an aquarium for a large clam hatchery (Knop 1994 c). It is intended to keep clams in large restaurants until their consumption, rendering the conservation of killed animals unnecessary. Many aquarists may object but clams for consumption should come from hatcheries if the natural resources are to be protected. The prohibition of clams to be served in restaurants would certainly not stop illegal collection, probably putting clams on the brink of extinction. Large amounts of cheap clams from hatcheries could render the trade with illegally caught animals unprofitable. Revenues from the sale of giant clams could then partly be used to repopulate devastated reefs. Therefore, it is clear that the propagation of clams in hatcheries for consumption plays an important role in the protection of these animals in the wild.

The aquarium construction I was asked for is intended to be used in tropical coastal areas. The material for construction must be cheap and easily available and laymen with no experience in the keeping of giant clams must be able to operate the device. It must therefore be easy to operate. I designed a double tank with a central filter chamber. The individual tanks are shallow to keep the necessary light intensity low and to facilitate manipulation of the clams. Both tanks can be filtered either separately or together through the central filter. The filter itself is divided into four chambers. Three of them are filled with porous material packed in netting. The material can be broken coral or live rocks. To prevent the formation of sludge in the biofilter, the first chamber should contain a layer of coarse plastic foam which must be cleaned regularly. To run the system in, "live rocks" from the reef are recommended as the first filter packing. The last chamber is holding the skimmer. A single pump is sufficient to produce the necessary flow of water to run the system and, in case of emergency, a second pump should be always present. The same measure is advisable for the membrane pump providing the air for the skimmer.

The filter chamber containing the broken corals must never be cleaned as it contains the bacteria for the biological filtration. This chamber is something like the "heart of the system". If it becomes necessary to clean them after prolonged periods of time, it normally takes three to four weeks for the bacteria to regenerate. The tanks do not contain any substrate and no permanent rockwork as decoration to inhibit the accumulation of debris, algae, parasites, or other infectious hazards. Only a few flat stones can be tolerated as they can easily be removed for cleaning. It is also easy to clean the walls from algae and the excrements of the animals can be removed with a hose. If cleaned in regular intervals, these tanks will always give a hygienic impression and the filter chamber will ensure biological filtration. Every one of the two tanks can be emptied according to the requirements and refilled either with natural or artificial seawater. If artificial water is to be used, it should ripen for one or two weeks before being connected with the filter. Otherwise, the chemically aggressive water may harm the bacteria of the filter system. For the same reason, temperature and salt concentration should be similar to the conditions in the filter.

The tanks are illuminated with fluorescent tubes. They are easily

▲ *Propagating of Acropora corals in captivity. Löbbecke-Aquazoo.   Photo: Rolf Hebbinghaus*

◄ *Cost effective aquaria are used to propagate a variety of sessile reef inhabitants.*

▼ *Tridacna derasa in an aquarium.*

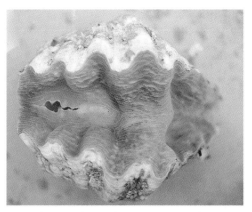

Tank for propagating Xenia at the Löbbecke-Aquazoo. Under optimal conditions, the Xenia population will double in two months time.
Photo: Rolf Hebbinghaus

Apart from Tridacna derasa, Hippopus hippopus is a favourite with specialty-restaurants.

Open tank system under construction. The tanks are connected with the nearby sea by pipelines.

obtainable and produce little heat. I recommend at least four tubes as long as one of the tanks (wet room fixtures). The light provided by this arrangement is sufficient to keep the clams healthy for at least four weeks. If the tanks are used alternating, the lamps can switched over from one tank to the other. Illumination with metal halide lamps would certainly be better for the well-being of the animals but the great heat production poses problems, especially in a tropical country if the tanks are not located in a fully airconditioned room. In case of failure of the aircondition, the heat production would pose a serious threat to the animals. Therefore, fluorescent tubes are recommended.

Of course a single tank could be used for the purpose but the possibility to clean out one tank and fill it up with fresh seawater while the other one is still in use helps to reduce the thread of infectious and parasitic diseases. While cleaning one tank, the animals, after having been cleaned in seawater with a brush to eliminate debris and parasites, can be kept in the other one. If using natural seawater, the artificial addition of trace elements and nutrients becomes superfluous if the water is changed in regular intervals not exceeding two to three months.

An important point is the regular monitoring of the salt concentration. With every clam removed, the water level will be lowered and must be replenished with natural seawater. Apart from that, however, evaporated water must be replaced by tap water to keep the salt concentration at the correct level. This should be done daily. Monitoring the salt concentration is best done by using a refractometer.

Open Aquarium Systems

Open aquarium systems which are connected to the sea are rarely found with aquarists. This rather simple way to run a seawater aquarium is mostly found in connection with public institutions like the aquarium in Monaco, Vancouver, Monterey Bay and many others. The method is also used when propagating marine organisms on a large scale. The most important advantage is that planktonic organisms and trace elements are always available. It is the only system in which organisms feeding on plankton without symbiotic algae can be kept alive over longer periods of time without developing signs of degeneration. An other advantage of open systems is that the water temperature is kept relatively stable. In the tropics I have experiences a rapid raise in temperature in such systems in case the water flow was stopped for various reasons. In no time the water temperature rose to over 30 degrees Celsius. If, however, the water flow is kept at an appropriate level, the temperature will hardly ever be higher than in the coastal water.

To save energy for illumination, often such systems are located in the open. In the dry season the tanks should be protected from strong sunlight, This is normally accomplished by using netting material.

The technical equipment necessary for such a system is rather simple. The water must be kept in motion either through aeration or water pumps. Often the flow produced by the pump supplying the fresh seawater is enough. The overflow catching the replenished water must be fed by surface water thus preventing accumulation of protein at the surface. If the tanks are without a roof, the overflow located near the surface will also remove rain water. Because of its lower specific weight it will float on top of the seawater and will thus be easily lead away without serious influence on the salt concentration. This is an important consideration in the tropics with its heavy rains at times. An other possibility to avoid excessive dilution of seawater by heavy rains are plastic roofs open at the sides. They can be used for open as well as closed systems. I have seen clam hatcheries using such roofes for cultivating the algae on which the Tridacna larvae were fed.

Open systems are technically much easier than closed systems and it pays for larger enterprises to be flexible in choosing a suitable site.

Export, Trade and Regulations

7

Like many other endangered animals, also the giant clams are protected by §1 of the Washington Species Protection Act (WA) which is enacted by the Convention on International Trade on Endangered Species of Wild Flora and Fauna (CITES). All exporters of the member countries need a valid export license (CITES clearance) to be able to legally export giant clams. This license provides a survey of the traded numbers of individuals for each species and controls the trade. The license is only issued by the appropriate authorities, if the export of the respective species does not endanger the naturally existing populations.

On the other hand, the member countries can only allow the import of endangered species, if a valid CITES-clearance is issued by the exporting country. If this clearance is not provided the customs authorities may confiscate the animals and hand them over to appropriate institutions to care for them, for instance, to a public aquarium. This is not always the best solution for the animals, as the aquaria in the zoological gardens usually do not have unlimited available space and may not have the necessary know-how or the installations for the very specific demands of some exotic species. Yet such a procedure is inevitable if one wants to make sure that the CITES regulations are enforced and that they are effective.

But although these regulations make a lot of sense, they cannot stop the giant clams from becoming extinct, as will be explained in the following chapter. Their only chance to survive, if there is any, is

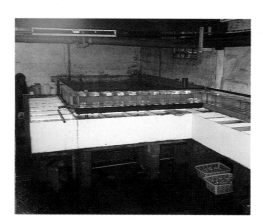

Packing station of a large Philippine aquarium fish exporter. The animals for export are kept and observed for 24 h in singular tanks.

Keeping clams in the same exporter. The export of these 1500 Tridacna crocea from the wild has been cleared by the Philippine authorities. Yet the condition was, that before export all secondary organisms living on the shells, like e.g. stony corals will be polished and brushed away, as these may not be exported due to species protection regulations.

by captive propagation. Therefore it is extremely important, that the aquarium trade with deals with Tridacnids as well as the supply for the deli-restaurants is not restrained by export limitations. The methods of captive propagation of giant clams are still very young, while the legal acts of many countries that regulate the import and export of the giant clams stem from a time, when captive propagation was not yet possible on a large scale and when the exported animals came from the wild. Obviously all the owners of giant clam hatcheries have a vital interest in selling their propagated clams. However, this is only possible if the authorities of the clam exporting and the importing countries adapt their regulations according to this new situation. Without any doubt the giant clams in the wild are endangered, but the propagated individuals are not at

The result of handicraft with shells is a souvenir from the "100 islands".

A curio shop in a Philippine province. If only the shells of cultivated animals would be used, these products would not be a problem for the protection of the species.

187

all endangered. Thus if the trade with these animals is limited or cut off, the hatcheries will not survive and the chance to save these species in their natural environments is gone. The propagation of the mussels is very costly and the production of the giant clams for restocking the reefs can only be financed, if at the same time a certain percentage of propagated clams is sold on the markets. The proceeds of this purchase has to be used not only for restocking the wild, but also has to be profitable for the farmers. Without any profit, the hatchery cannot survive in the long run. Already today, some giant clam hatcheries are close to going bankrupt, because the authorities in the respective countries refuse the export with the argument that the animals are endangered. These kind of restrictions do not help at all to save the species and the fact that in the same country the giant clams from the wild are sold in high numbers for consumption on many local markets, rises significant doubt whether species protection is really taken seriously. Each tourist can easily buy wild living giant clams on a market and eat them right on the spot. The empty shells of the individuals may, however, not be taken out of the country as a souvenir, as we are talking about an endangered species.

Certainly it makes sense in principle to forbid the trade with giant clam shells as well as with the products made from them to protect the mussels in their natural environment. These regulations are, however, not effective if the production of the curios is not forbidden as well, and if the law is not enforced. In fact the curio shops are loaded with these products, that are almost all made from shells that come from the wild. At the same time some of the giant clam hatcheries have enormous difficulties to recycle the shells of deceased, or in

deli-restaurants consumed propagated clams in an appropriate manner.

These examples show that the legal acts which regulate the trade with giant clams and which try to ensure their survival are in some countries no more in line with the changed circumstances and thus urgently need adaptive alterations to ensure that their effects do not turn into the opposite direction.

The urge of the giant clam hatchery farmers to represent their interests has lead to the foundation of the Association of Farmed Aquarium Clam Exporters, "AFACE", an international group of giant clam cultivators. When founded in Honolulu, Hawaii, this grouping formulated a number of regulations for export and trade with aquarium clams. For example, the very colourful clams should be made available to the aquarists, while the less colourful ones of the same species should be sold on the food market. The size of the clam, their colouration and their natural shape were supposed to be valuated by a specific key and categorized accordingly, so that all species have internationally comparable quality standards. This facilitates the pricing policy and prevents catastrophic price reductions amongst the giant clam producers. The ICLARM – Coastal Aquaculture Center in Honiara at the Solomon Islands that initiated AFACE, emphasizes that the purchase of giant clams cultivated at a farm, is an important contribution to the conservation of the natural stocks of giant clams all over the Pacific Ocean. Everybody who is busy with the research and propagation of these clams, knows how much the clam export for the aquarists supports the hatcheries and thus at the same time the restocking in the wild.

Until 1994, however, the export of cultivated giant clams to Europe was restricted to a handful of species. *T. crocea* was most commonly traded, whereas *T. squamosa, T. maxima,* or *T. derasa* were rare. The significantly larger growing *T. gigas* was never seen in Europe's aquarium trade and *H. hippopus* was even unknown. The sizes of the imported mussels varied between six and twelve, sometimes 15 centimeters. Larger or smaller animals were hardly found on the market. This is hard to understand, as the hatcheries should have had a pronounced interest in selling quickly. In particular with the slowly growing *T. crocea*, which was the main export aquarium clam, a very costly care lasting for years is needed before the juvenile clams reach sizes of six to twelve centimeters. Diseases and parasites are a major problem during these years and can significantly reduce the output and profit of a hatchery. Therefore the clam farmers should be interested in exporting animals of much smaller sizes, so that part of the growth to adult size would take place in the aquarium. In addition a small colony of ten or 15 tiny, differently coloured animals can be a very special treat and challenge for the aquarist.

If this concept would be realized, the hatchery could reduce the average time a clam spends within its tanks considerably. At the same time the percentage of successfully propagated animals in relation to the successfully fertilized eggs would increase by large. Also a stronger emphasis on the propagation of the faster growing species, mainly *T. gigas* and *T. derasa* could be profitable for the farmers. Altogether, a more variable supply of species of various sizes would promote the aquarium trade for giant clams.

Therefore, in 1994, I undertook an experiment to export cultivated clams of a variety of species at various stages of their development, to find out about the ability of these animals to adapt to the artificial environment of an aquarium. My experiments comprised the species *T. gigas, T. derasa, T. crocea* and *H. hippopus* of shell lengths varying between 15 millimeters and 15 centimeters. Their ages varied between three months and five years. I chosed the animals myself at a hatchery and accompanied them through all stations of their export. During packing different procedures were applied to test new possibilities for the transport of these animals.

The results of this export study (Knop 1994 a) showed that the clams can be adapted to aquarium conditions at considerably smaller sizes, than was so far the rule. Interestingly, the absolute lengths of their shells were not so important, as it turned out that the age of the individuals and thus their developmental stage was most critical. The growth rate varies a lot between species, yet the faster growing species like *T. gigas* and *T. derasa* are not at all more robust in transport or more adaptive to aquarium conditions, than the slowly growing *T. crocea*. A *T. gigas* may reach a shell length of 30 millimeters within half a year, but it is much less adaptive, than a one-year-old *T. crocea* with a shell length of only 20 millimeters. In summary, the study of transport and adaptation revealed that the animals need an age of at least twelve months to reach a developmental stage that allows them to adapt to aquarium conditions. At an age of three or six months they are not able to survive transport, although the faster growing species have already reached considerable sizes. Additionally, I could not find differences in animals of the

Careful choice of cultivated clams for the export study.

same age but of different sizes with regard to their ability to adapt, although deviation in size can reach 100 %. The survival of the clams was not size dependent and amongst the shells of deceased mussels there were those of the smallest as well as of the largest within a group of the same age.

In contrast to the experiments with rather young individuals, those with animals of the age of twelve to 18 months revealed much better results in all exported species. Therefore I proposed the regular export of the animals of this size in various clam hatcheries, so that, in future, they can supply the aquarium shops.

However, I think that the import of the cultivated clams is only sensible, if appropriate tanks for their quarantine can be provided. Also cultivated clams can be loaded with parasites and

Estimation of the volume of young clams before packing: Hippopus hippopus is seen here at the age of 15 months.

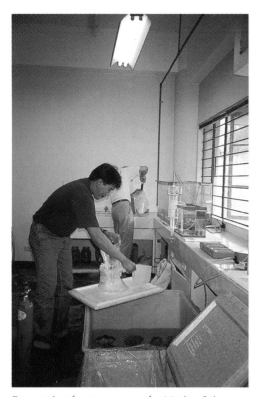

Preparation for transport at the Marine Science Institute of the University of the Philippines. The careful documentation of the applied packing method is an important prerequisite for sensible results in this comparative study.

The clams are packed very carefully.

Arrival in Germany. The transport of the animals and the observation of their ability to adapt revealed interesting results.

bacteria which would endanger entire stocks of giant clams. In addition, the correct and gentle adaptation of the animals to aquarium conditions needs a lot of experience. According to my opinion, it is absolutely vital that the freshly imported animals are kept for two to four weeks in quarantine before they are sold to pet shops or hobbyists, especially if a larger number of animals is traded which can infect each other. The quarantine tanks have to be appropriate for the needs of the giant clams and the animals have to be inspected carefully before they are traded. The cultivation of the giant clams in large numbers, should, however, not flood the market and make the clams become just "material", something that boosts the business until the aquarium trade is saturated. Obviously saturation of the market would not be in the interest of the hatcheries.

8 The Ecological Situation of the Giant Clams

The clam species *T. gigas*, *T. derasa*, and *H. porcellanus* used to be widely distributed in the Indonesian Ocean. In the meantime, however, their numbers were reduced considerably. In the Western part of Indonesia the species *T. gigas* has become completely extinct (Usher, 1984).

For the local people along the shores of Indonesia the *Tridacnidae* are one of their most important sources of income. The meat of the animals is part of their diet which is conserved by drying and mainly consumed by the shore dwellers. At the local markets this dried meat is traded for 1.50 US$ per kilogram. The shells of the animals are traditionally manufactured into wall decorations, mosaics, wash basins or ash trays, as well as used in the tile producing industry which is interested in these animals since the beginning of the sixties. Although the collectors of these industries mainly use the so-called fossil shells, which are the remains of deceased clams burrowed deep down in the debris and the litoral, dramatic damages to the reef are caused during the search for these "fossils". One to two metre long iron poles are used to dig and poke into the pebbles of the reef in the search of the shells (Pasaribu, 1988). The operation is usually handled from small about three metre long canoes. Recently some of the collectors are provided with snorkeling gear which improves the effectiveness of their search. Syárani (1987) examined the effects of the fossil search on the living stony corals in the reefs of the Islands of Karimun, Java and observed dramatic damage of the overall coral fauna. He concluded that the natural regeneration is not completed even in the course of many years. The damaged parts of the reef with the injured coral surface tend to make the water turbid and are often covered with algae that prevent the corals from recovering. This way the intact stony coral reefs turn into algal gardens which expand more and more and eventually suffocate even the last remains of corals.

The basic problem in Indonesia is the same as in all other countries. Use and exploitation of the clams are increasing, while the natural resources decrease. The smaller the stocks of natural tridacnas, the higher the prices. The higher the prices, the more frantic the search for them and even legal restrictions for the protection of the animals are simply ignored.

This is even more so, the less the existing regulations for their protection are executed by official authorities. For reasons to do with personnel shortage (due to a lack of financial sources) control and supervision takes place only very sporadically – if at all. Therefore regulations hardly discourage the poachers anywhere in Indonesia. The only exception is the capital of Indonesia, Jacarta, where governmental authorities for the protection of nature cooperate closely with the police force as well as local administrations. There it is shown that the effective execution of regulations is possible, at least in theory.

The pressure on the natural stocks of clams in other countries is not less. In Kiribati, for example, where fishermen harvest "te kima" (*T. gigas*), "te were" (*T. maxima*), "te were matai" (*T. crocea*) and "te neitoro" (*H. hippopus*) regularly, the natural resources drop drastically.

People hurry to collect and eat the animals before they are fully grown. This is due to an old superstition: local people are convinced, that the clams leave their shells when reaching their maximum length, turn into a ray and swim away. Also here the clams are an important part of the diet and not restricted to the coastal regions. The clam meat is rarely dried, but rather conserved in salt (pickled) and then transported in barrels to the capital of Kiribati, South-Tarawa. A more modern method of conservation with vinegar and "sour toddy", the fermented juice of the bark of the coconut palm, becomes more and more popular these days (Taniera, 1988).

▶

Some of the animals of which the shells are shown in the two photos here, were confiscated from poachers.

"Clam cemetary" in the Marine Science Institute of the University of the Philippines (UP MSI).
▼

▲
A curio shop in Asia.

The clam meat is consumed either raw, or after being cooked in the "Kiribati- oven", which is a fire place digged 50 centimetres into the ground and covered with mats.

The scales of the shells of Tridacna squamosa are cut off to achieve a smoother surface.
▼

The shells of *T. maxima* are utilized as mortar to crush leaves for the production of traditional medicines. The large shells of *T. gigas* are used for cooking, as a baby bath tub or as a troff for the pigs. More peculiar is a form of life conservation of clams in Kiribati. The male family members collect the clams in the reef, usually *T. gigas* and *H. hippopus* and store them in the sea next to their houses in shallow water. The animals stay there until being consumed.

In the Philippines, in contrast, the animals are taken out of the reefs immediately before consumption. Women and children collect the molluscs in the flat water regions, while the men also dive into deeper reef zones

One of the curios made from clam shells is this little "Santo Niño", the "Holy Child".

▲
These kinds of holy statues are usually placed in the families home altars.

Millions of marine creatures are manufactured into decorations and souvenirs by the curio industry.
▼

Sale of giant clams on an Asian market place. The large Tridacna squamosa are sold for 2 to 5 US$ per individual.

◄

In the morning the animals are offered alive, shortly after being harvested and often they are consumed raw right on the spot.

to get hold of the clams. In the early days, people would cut out the soft parts of the clam while still under water and leave the shells in the sea. For larger animals, this method was particularly handy, as it is extraordinarily hard to surface such a large animal with its heavy shells without a SCUBA-tank. This is the reason why you can frequently find empty pair of shells in the reef. However, since the shell manufacturing industry is developing more and more at the Philippines, since these shells are made into ash trays, salad bowls or even holy water fonts in masses and since they are broken or sawed into pieces to make table

The fisherman gets just a few Cents for a Tridacna crocea and about half a Dollar for a middle sized Tridacna maxima or Tridacna squamosa. This is all he earns for the day for a family of eight to ten people.

In a few years this picture will be gone from the fish markets of Asia. Four of the seven indigenous giant clam species are almost extinct.

decorations and jewellery, the Philippine clam divers try to take each and every possible shell out of the water.

In some regions of the Philippines, the clam meat is eaten raw. In the area of

Central Visayas, for instance it is know as "kilinaw". Only occasionally it is cooked in coconut oil. Earlier the fried meat was dried in the sun and thus conserved. Although the meat of the tridacnas plays only a minor role as a

Clams of the species T. crocea, T. maxima and T. squamosa were offered as well as H. hippopus.

Detail of a relatively undisturbed coral reef.

A reef after being fished with dynamite. In the centre of the explosion the corals are completely destroyed, while in the periphery they are turned over.

What remains is a limestone desert. Only the numerous skeletons of stony coral remind us of the colourful species richness that existed here for thousands of years. In many poor countries people live from the sea in this way. In the Philippines lots of coral reefs look like this.

Close-up of a reef destroyed by dynamite. The striking poverty of the local fishermen forces them to use this destructive method. "If we do it", they told me, "we do not have a future. But if we leave it, we do not have a present."

protein source in the diet of the Philippine people - it is only consumed when there is no fish available - still numerous clams from the Philippine reefs ended in a cooking pot. The reason therefore is the export of clam meat which occasionally had enormous dimensions, like for example, the sale of 1000 kg of frozen meat of *T. crocea* from Potillo Island in 1984. Due to the present restrictions these kinds of exports are no more legally possible, at least not, if the animals are taken from the wild. However, there is no doubt that the regulations are sometimes undergone and that life clams as well as their meat are smuggled out of the country.

This led to the situation that of the seven species of *Tridacna* indigenous to the Philippines, only three are still present: the smaller species *T. crocea, T. squamosa,* and *T. maxima* which are all attached to the ground with their byssal apparatus. The four species which live on the ground without being attached *T. gigas, T. derasa, H. hippopus* and *H.*

porcellanus are hardly found any more within the archipelago of more than 7000 islands. Only *H. hippopus* can occasionally be found, although extremely rarely, and the very few specimen may not be enough to ensure the survival of the species. Also the fact that the populations of the clams decrease roughly at the same rate as the population of the people on the shores increases stands for its own. For example, while in 1988 at the thinly populated Potillo Islands of the province of Quezon 3528 clams were counted per hectare (UP Marine Science Institute), and in the province of Palawan 3357 specimens (Silliman University Marine Lab.), surveys of these two universities record not more than 31 animals per hectare in Central Visayas and only 11 in West Pangasinan. Also the distribution of the animals within the reef zones is very interesting. In shallow water, the pressure from overfishing is particularly high, as the animals are easily accessible there. The deeper the water, the less the natural clam populations are disturbed, although even in deeper water only those specimens survived that are well hidden. If you look very carefully, you can find these clams below large table shaped stony coral colonies of the genus Acropora (Gomez and Alcala 1988).

In the Philippines, as well as anywhere, one became aware of this unfortunate development. But counter action takes more than just awareness in some scientists and a handful of politicians. One of the reasons for the wasteful exploitation on the reefs is thought to be that the sea is seen as a communal resource in the Philippines. Therefore nobody feels responsible for the health and the state of the reefs as everybody only views his own benefit – a regrettable but very human character. Doubtless, however, the incredible poverty of the local people is one of the major reasons for the destruction of the reef. Fishermen in the villages of the islands, which – without any infrastructure – depend entirely on fishing, have no choice but to take more drastic measures if the fish populations and the catch rates drop, to be able to support their families. Fishing with dynamite is a solid part of the daily struggle for survival and it is very unlikely that countries like the Philippines and other developing countries will be able to overcome the economic problems without foreign help. Without international help programmes and sufficient financial support for these projects the increasing impoverishment of these densely populated countries and the encroaching destruction of their reefs will not be stopped in the future.

To limit the damage, at least partially, and to find ways for the protection of

The left half of the picture shows an intact reef. To the right is an area destroyed by dynamite. This flat zone stretching over vast areas was once inhabited by the same beautiful coral formations. This shows how urgently a solution is needed for the problems of poverty in the Third World.

the coral reefs, some of the local politicians start to manage the resources of the reef in cooperation with developmental organisations. Certain families of the coastal population take up responsibility for a particular stretch of reef, kind of a contract to harvest in that area which is limited in time (Gomez and Alcala, 1988). The basic idea of this concept is to induce some form of responsibility in the "sea-farmer", and at the same time to achieve intensive control of the reefs. It is hoped that this policy may at least stop further destruction of the coral reefs.

There is some reason for hope here, as this policy shows that people are aware of the problem and that they think of solutions. However, like with the illegal ivory trade, time is running for the tropical coral reefs – and everywhere the giant clams which still exist in the wild seem to be the loosers in this race. Of course this is not only the fault of the locals in the respective countries. Responsibility for the present situation lies to a large extent with the international fishing vessels, which usually sail under Asian flags. How dramatic the destruction can be, caused by systematic catching expeditions for clams, will be demonstrated in the following with the Solomon Islands. This archipelago comprising of some 800 islands is situated close to Australia and belongs to the few places in the world where all clam species are still relatively widely distributed (except the two very rare species *T. tevoroa* and *H. porcellanus*). Like in other regions the local people along the shores reduce the clam populations by collecting and consuming them irrespective of their sizes. The only exception are the members of the church of the "Seventh-

Car tyres are used as settling substrate for corals. These constructions were used for an artificial reef in Dumaguete, Philippines.   *Photo: Marine Laboratory of the Silliman University*

Within a few years numerous marine organisms had settled on this artificial reef. I found impressive stony coral formation in 18.5 metres depth, together with huge sponges and hundreds of coral reef fish.

Photo: Marine Laboratory of the Silliman University

Also clams of various species settle on the artificial reefs. The picture shows a beautifully coloured specimen of the species Tridacna maxima.

Day Adventists", who do not eat these animals for religious reasons. This effects mainly the regions of Rennell, Bellona, West-Province and some regions of Malaita, as the major part of the population there is a member of this church (Govan, Nichols and Tafea, 1988). The clam meat that is not consumed locally is transported to the capital of Honiara, to be offered at the markets there at the same prices as fish, which is about 1 to 2 US$ per kilogramme.

In all reef zones, particularly in the vicinity of the more densely populated islands, the larger clam species *T. gigas* and *T. derasa* as well as *H. hippopus* have become extremely rare. This is, however, not only due to the catching activities of the locals, who construct clam gardens like the people in Kiribati, where the animals are kept until consumption. Because of their high meat contents or because of their precious shells, these three species are the primary targets of foreign fishing vessels, which emerge here more or less regularly despite catching restrictions and occasional controls. As early as in the middle of the seventies people harvested the shells of the clams at the eastern end of the Guadal Channel in Marau Sound, which was then legal, to provide the Japanese market with the clams. Some 25 tons of these shells were shipped to Japan then. Ten years later in 1983, experiments were undertaken with controlled clam collections. A Taiwanese ship called "Kao Tung 1" had received a license and undertook a catching expedition in the Marovo lagoon, where 1318 clams were harvested within 41 days. Also the coastal inhabitants handed over their catches and received 20 Cents per kilogramme of mantle meat and 40 Cents per kilogramme adductor muscle. A total of 1227 kilogrammes of adductor

Another construction made out of bamboo. These pyramids also serve as artificial settling substrate for corals.   *Photo: Marine Lab*

muscle meat was the result of this collection. Years later one could still see that the respective reef areas had recovered in no way. Yet, in June of the year 1983, the Taiwanese catching vessel "Shing Hong 3" was brought in close to "Indispensable Reef", and convicted of illegal mussel catching. It had 10210 kilogrammes of clam meat on bord (Govan, Nichols and Tafea, 1988).

Four years later the catching vessel "Her Cheng Fong 3" was brought in near Roncador Reef, that sailed under Taiwanese flag. It had more than a ton of adductor muscle meat on bord, which was party deep frozen and partly fresh. Experts estimated some 10000 specimens of killed tridacnas.

On the Fiji Islands the situation for the clams is not any better. In 1984, for example, the total amount of the legal commercial clam catch (only mantle and adductor muscle) reached 40 tons. Between 1981 and 1988, 210 tons of giant clam meat (mantle and adductor muscle) were taken from these reefs only for commercial purposes. The consumption of the local people and illegal catches are not included in this number. About half of these clams, roughly 120 tons were from the extremely endangered species *T. derasa*, some 120000 adult animals (Lewis, Adams and Ledua, 1988).

These examples demonstrate how difficult it will be to ensure the survival of these endangered species. Besides the above mentioned catching vessels, numerous other ships were brought in which had smaller amounts of clam meat on bord. This shows that governmental controls are executed to stop the illegal catchers. However, the examples also show that despite the

Philippine fishermen sorting out and drying their poor catch.

controls the clams cannot be saved from the catching vessels effectively.

The markets are regulated by supply and demand. A shortage in supply increases the price and causes pressure onto the suppliers. The rarer the clams become, the more an Asian "clam lover" will be prepared to pay for his delicacy. The higher this price, the more inventive as well as reckless the clam catchers will operate in the reefs. A continuous surveillance of the endangered reefs will always remain an illusion as it is far too costly. Regulations and controls by the authorities, however, will not solve the problem.

Governmental research and aid programmes to control and save the remaining stocks of *Tridacnidae* on the reefs are only successful, if they are wisely organised and based on solid financial grounds. Yet the financial side of these projects are the main problem

of the respective governments, as these are usually developing countries. An example for this is Tonga, where in 1978 John McKoy of the New Zealand Fisheries Research Division undertook a survey of the *Tridacna* populations. His study was by order of the government and lasted over four months. A total of 72 circumscribed control areas were examined using snorkeling gear in shallow water and SCUBA in deeper reef zones. When using SCUBA gear, the mussels were not only counted and identified, but also measured for size. The maximal shell length was recorded and in addition 125 animals were tagged to examine the growth rate.

The researchers were also active on land. The clam meat supply was surveyed at the local markets, as well as individuals measured and the gonads taken to the laboratory for further examination. The study revealed that already at this time, many of the reefs

A typical choice on a small fish market. The fish are usually offered live to demonstrate their freshness.

were emptied from tridacnas. Out of the four once indigenous clam species, only two were encountered regularly: *T. maxima* and *T. squamosa*. Single specimens of *T. derasa* were found only very occasionally and *H. hippopus* was not encountered at all during the whole duration of the study.

In 1987, about a decade later, the New Zealand Fisheries Research Division intended to undertake a comparative study, to be able to interpret the development. However, after having examined eight of the control areas the study came to a halt and it was decided to carry on at a later stage. The intended continuous control of the endangered reef stretches had also not yet started at that time.

Another activity to ensure the survival of the clam stocks was initiated at the same time by the Ministry of Land, Survey and Natural Resources: the

famous "Clam Circle Project". The basic idea was to collect the few *T. derasa* that still lived on the reef and to bring them together. Conservationists were fearful of the animals being unable to fertilize each other if they lived scattered over the reef, too far apart from each other. Therefore the animals were arranged in two circles, one inside the other; the inner circle comprised of 33 animals, the outer one of 66. It was hoped that the clams would reproduce and disperse again over the reef. Five of these circle colonies were established, one in the port of Nukuálofa and the other four in the port of Vaváu.

An unfortunate side effect of the experiment was that some of the animals from the circles were stolen by clam catchers. To replace them, other animals were taken from the reef and thus the species *T. derasa* became rarer and rarer in its natural habitat. The same kind of problems can be encountered in

205

Many people in the insular countries are entirely dependent on fishing.

many countries all over the world. In the beginning the financial aid as well as the support by foreign experts is provided. However, as soon as the experts leave, there is only poorly trained local personnel, that is not in a position to carry on with the project. Additionally often the financial sources for the projects run dry within a short time period. What is left is good will and a more or less stranded research and aid project.

Yet there is still hope for the clams through propagation in clam hatcheries. If it was possible to propagate the endangered species in sufficient numbers, the market could be satisfied with these animals. This is the only way how the prices of clam meat on the Asian markets could be lowered, so that the illegal clam catch is no more profitable. Thus the remaining stocks on the reefs would be effectively protected

against further exploitation. Then one could start to slowly re-establish the thinly populated reef zones and thus support natural propagation. The difficulties that arise from this method will be discussed in the following chapter.

Only a small percentage of the coral reefs in the Third World are still intact.

Captive Propagation

The enormous threat to which the clams are exposed almost worldwide makes the efforts for their conservation understandable. But not only the conservation of the species as such is a matter of concern, but also the production of protein as a food resource due to their rapid growth. Years ago already, a concept was developed comprising several phases of which the last phase provided the coastal inhabitants of tropical Third World countries with the possibility to farm clams. It was hoped that the people would have a source of income with such farms which work similar to algae farms. At the same time a large number of clams would be produced for the food market. Due to a higher supply of

A fully mature Tridacna gigas is hauled aboard with difficulty – here only for research. Later the animal is brought back to where it came from.

About two decades are necessary before a giant clam reaches such a size.
Photo: Marine Lab. of the Silliman University

The Marine Laboratory of the Silliman University in Dumaguete, Philippines has contributed a lot to the research of giant clam cultivation.

clam meat, the world market price would drop, which would in turn impede the illegally operating clam catching vessels, as their business is much less profitable. As a side effect one expected to be able to restock the clams in their natural environment and found new colonies which maintain themselves, reproduce and disperse. This way the final goal would be achieved to preserve the endangered clam species.

One of the organisations that played a

The tanks at the Marine Lab in Dumaguete.

Taking photos at the Marine Lab.

These tanks of the UP MSI are connected to the sea via pipelines

Small cultivation tanks at the UP MSI where the young clams spend the first couple of months of their lifes.

key role in the development of captive propagation was the Micronesian Mariculture Demonstration Center (MMDC) in Palau. In the seventies and eighties valuable information was collected here to develop artificial propagation of the *Tridacnidae* from a lab experiment to mass cultivation. Also the University of Papua New-Guinea (Motupore Island Research Center) was

Canvas lined cultivation tanks at the UP MSI. Here also, the researchers have achieved a lot in the last couple of years.

Choice of appropriate parental individuals of Tridacna gigas for cultivation.

Photo: ICLARM-CAC, Honiara

active in this field, although less comprehensively. In the early eighties Australia appeared on this front, firstly in form of a smaller governmental research project which was carried out by the University of New South Wales. Later Australia came up with a larger and comprehensive project granted by the Australian Center for International Agricultural Research (ACIAR) and carried out by the James Cook University in North Queensland. The first step of this big ACIAR research project comprised Australia, the Philippines, Papua New-Guinea and Fiji. The second step at the end of the eighties additionally included Kiribati, the Cook Islands, Tonga and Tuvalu.

The idea spread around the world. The International Center for Living Aquatic Resources Management (ICLARM) founded the Coastal Aquaculture Center at the Solomon Islands close to Honiara, where besides a few other marine organisms, mainly giant clams are cultivated. More hatcheries emerged at Tonga, the Cook Islands and mainly in Micronesia. A busy exchange of information began worldwide amongst biologists who were involved in the research of giant clams. Everybody made his piece of information available to the others to facilitate the research and the captive propagation particularly effectively. Around the globe a group of scientists found together who has a special interest in these fascinating molluscs and therefore worked on them intensively: the so-called "clam family". The ICLARM Coastal Aquaculture Center soon published a small newsletter entitled "Clamlines", which was sent to the members of the clam family all over the world twice a year. In this simple brochure new findings in science were published and thus made available to the other biologists interested in this field. At the same time the ICLARM started a data base where all data of all clam experts were collected. With the help of this data base it was possible for every scientist to contact other experts quickly and easily. In general an atmosphere developed amongst the clam specialists which was characterized by pioneering emphasis, openness and a deep sympathy for the *Tridacnidae.* Hostility and competition were unknown amongst this global network of biologists. I myself was lucky enough to experience this harmonic atmosphere in many places and I was especially impressed by this very effective way of cooperation.

While in the early stages one was mainly concerned with food production and restocking of clams in the reef, in the nineties the aquarium trade was discovered. It turned out that the aquarist's market was considerably bigger than originally viewed, that the hobbyists were prepared to pay high prices and that this was a potential to support commercial clam hatcheries financially. Particularly in the beginning it is important to be able to sell the propagated clams, to surmount the initial investments. This was a crucial point for many farms as it takes a few years before the first adult clams can be sold. Up to that point there is no income while investment continues. This is relatively easy with the large and fast growing clam species like *T. gigas.* Certain markets, however, only import the small species *T. crocea*, like e.g. Japan, where this species is viewed as a delicacy. It was difficult to produce sufficient amounts of large enough

The chosen individuals are carried to the hatchery on land.   Photo: ICLARM-CAC, Honiara

211

Preparations for an experiment at the Marine Science Institute of the Philippine province of Pangasinan.

animals as this smallest species is at the same time the one which grows slowest. While this food market was mainly interested in adult *T. crocea*, the clam hatcheries could sell the subadult specimens which could be produced in a much shorter time in large numbers to the aquarium trade. Thus the liquidity of young hatcheries was considerably increased and the time it takes to achieve profits shortened.

The research on captive propagation was, however, not an easy task, despite the very effective work of the scientists and drawbacks and defeats were inevitable. In the following I will briefly describe what happened in the years between 1985 and 1986, when the species *T. tevoroa* was discovered. It started in August 1985 during a commercial clam harvest at the Eastern Fiji Islands. Various species were collected, but 95% of the harvest comprised of *T. derasa*, which is the most common species in this area. The collectors noticed very soon, that some of the clams differed significantly from the others. They had a much darker mantle with a warty appearance and much sharper shell rims. The two biologists A.D. Lewis and and E. Ledua thought initially of a local variant of the species *T. derasa*, which could have evolved due to special environmental conditions. This is in fact normally the origin of a new species, which then develops independently when prevented to interbreed with the original species by physical barriers.

Surveys in the local population revealed later, that this clam has been known under the name of "devil clam" ("tevoro clam") since long, while it was unknown in science. This is not at all unusual, as three years earlier, in 1982, the biologist J. Rosewater scientifically described the species *H. porcellanus* for the first time, a clam long since known under various names amongst the local people of many tropical countries.

In January 1986, when the biologists Lewis and Ledua discovered such a clam amongst hundreds of *T. derasa* about 130 nautical miles north of Vatoa, they decided to submit a thorough

Subadult giant clams (Tridacna gigas) at the UP MSI. More than 4000 specimens were bred here in the course of a few years.

examination. As they had no facility at sea to transport this surprising an unexpected discovery, the clams were killed and their soft parts deep frozen. Unfortunately this tissue was lost during a later flood at the research center.

People who have become curious, now tried to find these unusual clams systematically, to be able to examine them. After a search in June of the same year in the vicinity of of Vatoa remained unsuccessful, a reward of 50 Fiji dollars for each live specimen of such a clam was allowed. More than 300 animals were delivered by collectors. In six cases the catchers reported of a discovery of the "devil clam", but in only three cases this could be verified. The animals were deposited on the reef and transported live to the research station. Two of the clams did not survive the transport, the third one was in a very bad condition, when it was deposited in the flat water zone close to the shore. Nevertheless it was brought to the station in Makogai. During transport this third specimen died and its soft tissues were conserved in formalin.

Gonadal biopsy of the species Hippopus porcellanus in the Marine Lab of the Silliman University.

Requisites for gonadal biopsy.

In January 1988, the team of researchers visited Vatoa for a third time, after it was told that four more "devil clams" had been discovered. The thorough search in the respective part of the reef brought a fifth specimen from 20 metres depth to daylight. This animal showed a marked sensitivity to light, when brought to the surface. One of the five clams at this site died right on the spot, the other four were brought very carefully in a 500 litre tank to the Makogai Institute which was 220 nautical miles away. Although the animals after one week seemed to be still rather healthy they all deceased within the second week.

Injections of serotonin trigger the release of gonadal products. Here the rare Hippopus porcellanus is injected (UP MSI).

A number of specimens of Hippopus porcellanus are exposed to the sun to trigger egg release by an increase in temperature (UP MSI).

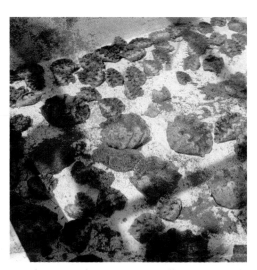

A cultivation of Hippopus porcellanus in March 1990.
Photo: Marine Lab of the Silliman University

The sea water is filtered to remove planktonic organisms, which may be dangerous to the eggs.

The reasons for this mortality remained unclear for long. However, it was assumed that the clams which had almost all been collected in deeper water (15 to 20 m) did not tolerate the stronger sunlight in the three metres deep coastal zones. I believe that the adaptive mechanisms of these clams were overstressed by the sudden increase of light intensity (see "central bleaching" chapter pathology)

In the following period when the news of the "devil clams" spread more and more, reports of discoveries became more frequent. Within the same year five specimens were found near the island of Komo and four more animals near the island of Namuka. Ten more individuals were brought to Makogai and this time kept in much deeper water. With them the first experiments for propagation were undertaken. In the meantime successful propagation was

reported, but the young were lost during a tidal wave. Despite all the drawbacks, that went together with the loss of representatives of this rare clam species, it was possible to collect more and more information about it. Maybe one day captive propagation will be possible and then the prerequisites are given to save this highly endangered species from extinction.

The Hatchery

The propagation of giant clams has two phases: the "hatchery phase" and the "ocean nursery phase". The first phase of which the name derives from the word "hatching" comprises the artificial reproduction, hatching of larvae and their growth up to a size of two or three centimeters. In the second phase the animals live in a sort of

Tanks for planktonic micro-algae for feeding the giant clam larvae.

To protect the algae from direct sunlight the drums are covered with dark nets.

Larva container; for one week the fertilized eggs are kept in this container.

underwater cage in the flat water zone to grow safely to the intended size. This second phase is consequently called "nursery".

At the beginning of the hatchery phase the choice of appropriate parents is needed. These animals can stem from the wild, or they are adults from the farm itself. The only important thing is the female maturity of the clams, so that not only the sperm, but also ripe eggs can be released. In general the production of fertile eggs is much more difficult than the production of sperm, because the release of eggs depends on the presence of fully mature or at least almost fully mature individuals. In addition, not every adult individual is prepared to release eggs at any time. Therefore, as a rule, one tries to bring together a number of adult animals for the release of the gonadal products. In the smaller species this is not very difficult. In *T. crocea,* for example, a sufficient number of adult individuals are usually continuously available. This is not always the case in the larger species *T. maxima* and *T. squamosa.* In *T. derasa* and even more so in *T. gigas* this can be an insurmountable problem,

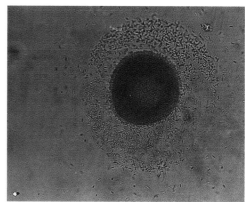

Polyspermia; this egg is surrounded by hundreds of sperm cells and will not develop.

Technical diskussion at the breeding tank. This round concrete tank contains some 25,000 juvenile clams of Tridacna gigas which are 3 months old and have a shell length of about 15 mm.

as these species are almost extinct in certain countries. As mentioned in the previous chapter, the entire Philippines, for example, dot not have a single adult specimen of the large *T. gigas* left. A further difficulty is, that particularly the adult individuals of this large mollusc species cannot be transported around the world very easily and reproduce elsewhere. Thus only those clams can be propagated that are available in the respective country.

Already a few years ago, however, people started to send the fertilized eggs or larvae to other breeding stations and thus enabled the scientists there to work with those species. The Marine Science Institute of the Philippine Province of Pangasinan has, however, so far not been successful in artificial egg release with *T. gigas* as all the available individuals are still too young. All the

larvae of these roughly 4000 subadult *T. gigas* stem from other research institutions, in particular from the

Blocks of artificial cement substrates with juvenile clams of Tridacna crocea.

A round tank with cement substrate blocks at the Marine Lab. in Dumaguete.

ICLARM Coastal Aquaculture Center, which has a large stock of adult specimens of this species. Yet my friends Bert, Eric and Hilly-Ann have in the meantime a lot of experience with the raising of the clams and it is only a

Comparative studies have tried different corrugations in the cement substrate to induce the settlement of as many juvenile clams as possible.

matter of time, until the *T. gigas* bred there will reach their female maturity and the intensive efforts to "spawn" them will be rewarded. This would produce their first F1-generation, which is the first generation offspring. In other species, the scientists of Pangasinan have already reached this stage.

If there are neither cultivated parental animals, nor fertilized eggs or larvae available, the breeding stock animals have to be collected in the wild. When choosing these individuals, it is not only important to consider their health status and the attractiveness of their colouration, but also to choose an animals with sufficient shell length, because only adults have a chance to reproduce successfully in captivity. The breeding stock animals are collected with great care and brought aboard a boat. In large specimens of *T. gigas* this can be very hard and special technical installations – like a set of pulleys – is necessary. The care for the animals aboard is not complicated as long as the trip lasts less than two hours. While travelling the clams should not be stored upright as this position induces a strong pulling stress to certain tissues due to changed gravity conditions; the soft tissues of the animals are considerably heavier, when out of the water. If the clam is laid on the side, covered with a wet sheet to provide humidity and a shade against the sun, and watered with sea water every 15 minutes, the animal can survive two hours of transport without difficulty.

Occasionally a gonadal biopsy is carried out, which is the removal of a tissue sample from the gonads, to choose the breeding stock animals. This tricky method provides a more precise picture of the state of maturity of the gonadal tissue. If ripe eggs are present, it is very likely that some of them will be found in the tissue sample. In this case the respective animals will be chosen for breeding. However, this method is not infallible, as only a single spot of the gonads is examined. Usually it is the tip of the organ in the front of the animal which is relatively easy to reach with the biopsy needle. Unfortunately this is the region which fills last with ripe egg-cells. Thus it is not unlikely to get a "false-negative" finding from the examination, which is, that the biopsy does not show eggs, although the gonads are full of them at the other end, directly beneath the kidneys. Also the gonadal biopsy is risky for the mussel. An inexperienced person can easily injure the hart, the kidneys or other organs of the animal by applying a wrong angle of puncture or by entering the needle too deeply. Therefore this examination can only be carried out by

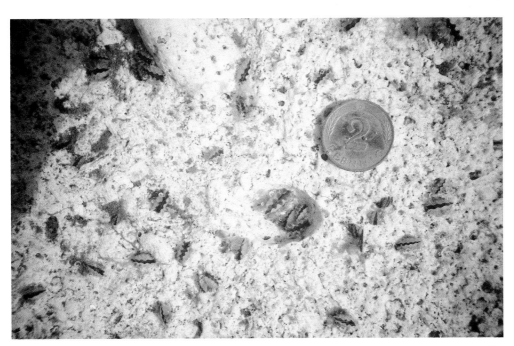

Tridacna crocea at the age of a few months. At this stage of development a high variation in growth is already visible (A German coin of about 2 cm in diameter serves as a reference).

The crevice at the bottom of the pebble is a preferred place for the clams.

Juvenile Tridacna gigas at the age of a few months.

well trained personnel and only if there are enough adult animals available from which can be chosen for a breeding experiment.

Once the breeding stock is chosen, one has to think about the appropriate stocking. This concerns on the one hand the continuous stocking of the animals in the shallow water zone, as they will be used repeatedly for breeding experiments over a longer period of time. On the other hand it concerns the short time stocking during the experiments themselves. The latter is usually done in so-called "open systems", flat stocking tanks, which are connected to the sea by pipelines. To be on the save side, the outside of the shells are cleaned off all secondary inhabitants to remove potential parasites.

To prepare the animals optimally for reproduction the water temperature is increased an additional food is offered consisting usually of phytoplankton. By these measures, which are also applied in commercial oyster farms, the metabolism of the animals is improved and the willingness to reproduce enhanced.

There are different methods to trigger the release of eggs and sperm. The release can be spontaneous without any recognizable stimulus. It can also be a reaction to drastic changes in the environment, kind of like shear despair in a hopeless situation, or, it can be induced by certain substances. There are two methods for this procedure called "induced spawning", which I will explain later. For now, lets go back to spontaneous egg and sperm release. To trigger it, the clams have to be exposed to artificial stress. In the beginning of this procedure, the tank water is raised to 33 degrees for a few hours. In an open system which runs in the heat of the tropics, it is sufficient to just interrupt the water circulation and the sun will quickly heat the water. After this phase the animals are taken out of the water and exposed to direct sunlight for 30 to 60 minutes. Mechanical stimulation by poking which is practised in some places is according to my opinion questionable, although it is sometimes successful. The pulling forces on certain parts of the tissue by the dangling soft tissues are anyway enormous when the clam is out of the water. Poking can induce such strong forces that the sensible tissue is severely

At this stage of development algae are the greatest danger for the juvenile clams.

Regular brushing is vital for the survival of these algae covered clams.

injured. Therefore this method is counter indicated, particularly in very rare and precious species, like the adult *H. porcellanus* which is shown in the figure. After being exposed to direct sunlight the animals are brought back into the water. After a short while, in many cases, one of the animals will release sperm. It is much more likely that an individual releases sperm than

Mortality is high in clams of this size and developmental stage.

As soon as the clams have settled on the artificial settling substrates, they can easily be moved from one tank to another.   *Photo: Marine Lab*

Tridacna gigas at the age of 3 months. A singular larger clam is introduced into the tank as an "indicator". During the first couple of weeks while the other clams are still too small to be seen, it indicates the water quality.

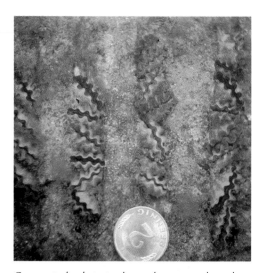

Corrugated substrates have shown good results. This photo shows Tridacna crocea at the age of 8 months growing in rows.

1.500 Tridacna crocea are being introduced to the tank after transport from UP MSI in Pangasinan to the Marine Lab in Dumaguete.

eggs, but the release of sperm can induce the rhythmical contractions in other clams which stimulates the gonadal tissue and eventually triggers the release of eggs as well. Released sperm is caught in a cap and stored separately. Yet the success of the whole procedure depends on the release of ripe eggs.

If there is no spontaneous release of eggs, there are, as has been mentioned, two methods to artificially induce it. First, the gonadal tissue of an adult clam can be put into the water in macerated form to provide the animals with the trigger substance "SIS" (see chapter "Reproduction"). Second, the hormone Serotonin can be administered in small doses that has to be injected into the gonadal tissue for this purpose, similar to the gonadal biopsy procedure. This

hormone influences the bodily tissues of the clams as a neurotransmitter: it enhances the transmission of electrical impulses and causes the reflectory

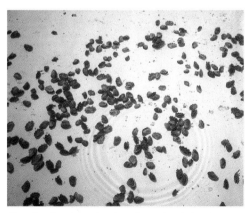

Shortly after introduction the animals begin to settle on the bottom of the tank (age: 17 months).

223

contraction of the gonads. This usually induces sperm release within one to five minutes and in case egg release after 20 to 30 minutes. The first method requires the killing of an adult animal, the second method beares the risk of injuring a breeding stock animal with the needle through incorrect injection. While the application of macerated gonadal tissue has been proposed in early research (Wada, 1954; La Barbera, 1975; Jameson, 1976), the Serotonin-method is a more recently published procedure (Braley, 1985, 1986, Crawford et al. 1986; Alcazar et al. 1986) and now it is most common all over the world. The Philippine scientist Sally Alcazar of the Marine Laboratory of the Silliman University made it possible to breed the very rare species *H. porcellanus* successfully by injections of serotonin. In the meantime people work on the F1-generation of those clams.

Tridacna crocea at the age of 8 months (A German coin of about 2 cm in diameter serves as a reference).

Occasionally the tissue from the biopsy can be used to induce spawning, if it contains ripe egg cells; like the macerated gonadal tissue from the

Tridacna crocea at the age of 18 months.

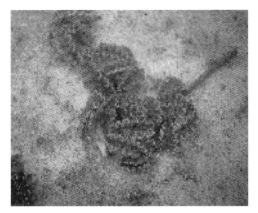

Also as a juvenile Tridacna derasa is amongst the most attractive clams. The animals seen here are 5 months old.

Released sperm is caught and stored in a separate container, as mentioned above, for instance in a clean and if possible sterile bucket. If the sperm would remain in the tank, the eggs would suffer from a so-called polyspermia. In contrast to higher developed animals the fertilized eggs of the tridacnids cannot hinder additional sperm cells from entering, which blocks the further development of the egg. To prevent this, we have to separate the sperm to ensure that roughly the same amount of sperm cells is poured over the eggs which corresponds the concentration of in the sea during spawning. I will come back to this in detail later in the text.

sacrificed animal it is poured into the incurrent syphon of another clam. Superfluous gonadal tissue with egg cells can be deep frozen for later use.

If the catching of the sperm causes a problem, the procedure can be simplified by translocating the clams into a tank with clean water. Before

A group of small juvenile clams can be very pretty in a sea water aquarium.

The clams prefer the crevice on the bottom of this artificial substrate.

Larger animals are tagged with numbers in order to record their individual development.

translocation, the animals are, however, cleaned with sterile sea water to remove all sperm that might have attached to the shells. The original tank is now a sperm reservoir. This procedure has to be repeated, if the clams release sperm in the new tank again. Yet, if they release eggs they can be collected. This is usually done with a plastic bag, which is positioned just below water surface above the excurrent syphon of the clam, or we collect the eggs with a plastic container from the water surface. I advise to try a funnel, at the end of which a tube is attached. The funnel can be positioned above the excurrent syphon while the end of the (wide) tube is within a container. Water is sucked into the tube but the end is blocked off until the clam starts with its contractions which is preceded by a water intake. As soon as the contractions start, the tube end is opened and the water runs into the container. This way most of the eggs can be caught. However, the density of eggs in the container may not be too high, as this would cause a shortage of oxygen. This would harm the eggs considerably and higher mortality rates with a higher intoxication by micro-organisms in the tank water would be the consequence.

Cages protect the clams against predators.

For fertilization of the eggs 30 to 60 ml of the sperm water are poured into the container with the eggs. The volume of this container should be at least 30 litres, to prevent polyspermia. If

Juvenile clams before their transfer to the cages in the sea.

227

necessary, the amount of sperm water can be reduced. After this measure we leave the container for about an hour in a quiet place. Then the water should be exchanged for fresh sterile sea water to clean the milieu from surplus sperm and organic matter. This is easily done with the help of gravity. As the container has been left in a quiet place, the cells have settled on its bottom which can now be sucked via a tube into another container, ideally through a sieve of 200 microns, so that coarse cell material and dirt of all kinds is removed.

If a microscope is available, the first cleavage phases can be observed already two hours after fertilization. At that time, most of the eggs should be in the two-cell cleavage phase, while some are already starting the next cleavage.

This way we can estimate the rate of fertilization, which is very important, because, if the rate is only 50% we have to be aware that very soon we will have a large number of deceased and fungus infected eggs.

Counting the eggs is also very easy with a microscope. This is important to be able to manage the density in the tank as we cannot filter or skim the water in this container. If the density is too high, the water quality of the tank will quickly drop and as as soon as a certain amount of egg cells has died the milieu can become toxic. Therefore a density of one to two egg cells per milliliter should not be exceeded. For counting the contents of the container should be mixed carefully but thoroughly to evenly distribute the cells.

The clams need regular care to grow well.

In the floating cages, the juvenile clams are protected against most predators
Photo: ICLARM-CAC, Honiara

Then we take a number of samples of 0.5 ml which are counted in a grid chamber (Sedgewick-grid chamber) under the microscope.

Besides the low density, hygiene is of importance in our larva tank. All material in contact with this water have to be as sterile as possible. This is of course also true for our hands, particularly if we have handled sea water before. Numerous micro-organisms which are harmless in a reef aquarium could otherwise infect our container and kill the sensible larva. Invading bacteria and fungi would reproduce enormously and the dying larva would quickly intoxicate the milieu. In the professionally operating hatcheries disinfection of hands and

The algae growing at the surface of the cages have to be cleaned off regularly.

229

Twice weekly these animals are controlled by SCUBA-divers.

other material are therefore usually strictly controlled and antibiotics are used in addition to prevent the dispersal of the bacteria. The most common product is Streptomycin (1 g/ 100 litre water) which is effective against both gram-positive and gram-negative bacteria ("Gram"-colouration is a procedure in microscopy). Although the effectiveness of this treatment can be enhanced by combining it with other antibiotics like Neomycin and Penicillin, it is considerably more difficult to find the correct dose. It might be possible to also use Chloramphenicol, but this antibiotic can be hazardous to humans (in a small percentage of the population Chloramphenicol can cause cancer) and therefore I would not recommend to fiddle around with this substance.

If no antibiotics are available, the perfect hygiene is even more important. We should make sure that the bottom of our tank is clean and try to avoid germs from invading. The water is aerated softly with rather coarse bubbles to ensure that the eggs are moved around. Temperature should vary between 26 and 30 degrees Celsius. It is however advisable to keep the larva at 26 to 27 degrees Celsius, as the bacteria reproduce quicker, the higher the

Tridacna derasa is moved to the cages in the sea.

temperature. The propagation of the bacteria can thus be retarded. As a consequence also the larvae are delayed in their development, but this does not do any harm.

Twelve to 18 hours after fertilization the free-swimming larvae hatch from the cell-clusters. Still, there might be some unfertilized eggs which are not developing and which produce oxygen consuming decay processes and the propagation of micro-organisms. To avoid this, the unfertilized eggs have to be removed. However, to make sure that we are not removing any fertilized eggs by mistake, we should wait 24 hours (after fertilization) during which the fertilized eggs will all have developed into free-swimming larvae. At that stage we interrupt aeration for 20 to 30 minutes and let the dead material sink to the ground from where we can suck it off with a tube. It is important not to cause turbulences during this procedure. Afterwards the water has to be replaced by fresh, sterile sea water and in case the correct dose of the antibiotic has to be added to keep the concentration of the drug in the water constant.

After 48 hours the larvae should be transferred into another tank. For the transfer, the aeration has to be switched off and the water is sucked through a hose into another tank. The last three or four centimeters in the tank should not be transferred as this layer comprises most of the infected or dead larvae. The density in the new tank should also not exceed one or two larvae per milliliter water. Many hatcheries work with much higher densities, but they can provide sterile conditions and they always use antibiotic treatments.

The Metamorphosis

After about one week, the larvae begin to settle on a substrate. To avoid them to settle on the bottom of the tank, we should provide them with appropriate limestone substrates. Experience has shown that the larvae prefer natural limestone or a thoroughly watered cement plate to smooth glass of plastic surfaces. This is due to the coarse structure of the surface which provides better anchoring; on such a surface

231

many more larvae settle than on a smooth surface. But this is not the only reason why we should provide the larvae with limestone. The cleaning of the tank is, for example, much easier, if the larvae are not attached to the tank bottom, but to removable substrate plates.

The number of free-swimming larvae will decrease with time. To see the larvae we use a torch and shine into the tank during the night. In the ray of light the free-swimming larvae can be recognized as floating illuminated dots. When shining into the tank, we should also examine other waste material like bacteria which can be seen as threads which float in the water column. These have to be removed carefully as otherwise the entire brood can be lost. If they appear in great numbers, the larvae have to be sucked through a tube into another tank.

The yolk sack with which the larvae hatch ensures their nutrition during the first couple of days. After it is consumed the larva has to start to feed plankton. As it does not have symbiotic algae at this stage in its life, feeding is particularly important. It is possible in principle to raise some of the larvae without feeding, but the survival rate is much higher if the brood is fed. As food cultivated unicellular algae of sizes between two and ten micrometers are appropriate. The density of food particles, however, has to be rather high, yet we have to make sure that the water quality does not suffer from rotting food.

Another possible food type is yeast which is, however, controversial in giant clam breeding. Fresh yeast or dried yeast turns the water into a milky soup, which is cleared by the larvae within a few hours. Pure and liquid coral food on yeast basis can also be tried, if it does

Cultivated clams of the species Tridacna derasa with a shell length of 15 to 20 cm.

not contain larger floating material. The professionals apply in addition a number of other food types, which partly achieve considerably higher survival rates. Here I will restrict myself to the already mentioned food types, because the others are either still in an experimental stage, or because they require such a high productive effort, that they are not available for the hobbyist.

A few days after the larvae have settled the process of metamorphosis starts. The larvae change their whole physiology completely and turn into a little clam. The tissue of the mantle between the incurrent and the excurrent syphon starts to proliferate to make room for for the symbiotic algae. In the natural habitat of the clam, the symbiotic algae are taken up from the water together with the food by the gills. To supply the animals with these algae in a tank we administer them instead of planktonic food. The stage of development for this application is not

Tridacna crocea with shell lengths of about 8 cm.

yet very clear. Therefore it is better to start early with the application, already on the third day after fertilization and four more times on every second day.

The production of this algae solution is not too simple. We start with a piece of the mantle of a conspecific giant clam. It can be obtained from the parental animal, by blocking the closing of the shell with a wooden stick. The mantle piece is taken into a pair of forceps and cut off with a sharp scalpel. If this injury is not too close to the pallial line, the animal can regenerate it as long as the keeping conditions are good. The loss of tissue and the injury are similar to an injury caused by a predator, a fish, for example, which occasionally bites off a piece of mantle. However, if the clam is in a bad shape, or if the tank conditions are not optimal, the cut can have drastic consequences.

To get the symbiotic algae out, the piece of tissue is washed in freshwater and finely macerated. This is best done with a razor blade which is scraped over the tissue. The brown solution resulting from this procedure is filtered through a net of 28 - 55 micrometers mesh size which holds back the larger cell clusters. Afterwards the solution is centrifuged to get a more concentrated algae solution. The bottom layer is then washed off with sea water and fed to the larvae.

If the larvae or the juvenile clams manage to install the algae into their mantles, they have access to an additional source of food and thus can grow faster. The shells also grow quickly given there are enough calcium and carbonate ions available.

In the hatchery these young clams spend the first six months in flat breeding tanks, the so-called "land - nursery" which is kind of a protected care station under constant observation. The water level in these tanks varies between 30 and 50 centimeters. Cement blocks with corrugations and crevices are provided as substrate for the juvenile

Tridacna gigas at the age of 41 months with shell lengths between 25 and 35 cm.

clams to settle. The animals prefer crevices to hide their shells in them. If on a smooth surface, they attach to their conspecifics which normally will result in groups of clams of various sizes.

The young clams need constant care and a shade provided by nets that protects them from direct sunlight. The growth of filamentous algae which is normally controlled by algae eating fish in the reef has to be controlled with brushes in the hatchery, as otherwise the mantles of the clams are covered with algae in a very short time. By fertilizing the water artificially with nitrates and phosphates the growth of the clams can be enhanced significantly. This has been shown in a study by Carmen Belda, but the subsequent blow in algae growth also increases the animals' mortality. The introduction of algae eating fish has shown questionable results. Their excretions impair the water quality and

lead to decay processes if not removed carefully every day.

After the first six months of their lives (a duration that varies from farm to farm) the young clams move to the "ocean nursery". They are transferred to special breeding cages in shallow water (about two to four metres depth). The mesh size of these cages exclude larger predators, but allow sunlight to penetrate. The clams are first introduced into the cages on small trays. Later when their shell lengths have increased, they also attach directly to the bottom of the cage. The cages have to be controlled for parasites and cleaned from the algae covers on a regular basis. When diving on one of those farms, I could occasionally see cages of which the surface was almost completely covered with algae. This can happen very quickly and therefore the cages have to be supervised by diving personnel usually twice weekly, no matter if wind or rain make the trips difficult. To hinder bottom dwelling predators like snails or crustaceans from entering the cages, they are often put on stilts which makes them resemble tables. Furthermore the legs of these tables are provided with funnel like installations which are hard to pass by the snails and crabs. These kind of constructions are used, for instance, by the Philippine Marine Science Institute UP MSI. An alternative to protect the young clams against predators are the floating cages, which are used by the Coastal Aquaculture Centre in Honiara. These cages are an almost perfect protection against bottom dwelling organisms. Unfortunately not all parasites that are harmful to the giant clams live on the bottom. This was experienced painfully by the researchers of the CAC in Honiara when a flatworm pest (described in the chapter on diseases) broke out. These floating planaria entered the cages without difficulty and

propagated there extremely quickly due to the lack of predators.

With increasing shell length the clams become more and more resistant against parasites and reach their last stage in breeding. Living freely on the sea floor, they grow to the intended size in a number of years until they are harvested or used for restocking in the coral reef. The larger species have hardly any predators at this age, so that mortality is rather low under good living conditions. However, the clam colonies have to be checked after every thunderstorm, because the turbulences can turn the clams over or cover them in sand. In this case the divers have to act quickly as otherwise some of the animals can be lost.

If the colony can be successfully protected against thunderstorms and all sorts of other dangers, some of the individuals will be living one day in

After 5 years these animals have already reached an impressive size.

their natural habitat. They will habituate to their natural yet new surroundings, live there for a number of decades and eventually reproduce and have offspring. Thus, these cultivated giant clams have done their share to ensure the survival of the species over a critical period of human exploitation and they will continue their evolution.

It was a long time before the cultivation of the giant clams became possible and thus the survival of this unique group of animals could be ensured. The enormous efforts provided worldwide by many institutions and the great achievements by numerous individuals can partly be explained rationally. The benefits of cultivation are obvious: on the one hand the endangered species have a chance to survive, on the other hand the enormous growth rate of the giant clams can be used for protein production. This could be an important aspect considering the increasing human population on earth and the diminishing food resources that goes with it. Nevertheless, I believe that besides all these logical considerations, there is another reason for the intensive efforts for the cultivation of the giant clams. Wherever I met the members of the "clam family", I was fascinated by their emotional relationship to their clams and by the respect they had for these animals, whether they were juveniles of a few millimeters or huge adult specimens. Maybe this respect, along with the cultivation and the restocking in natural habitat is a way of paying back to the clams at least a small portion of what man has done to them in the course of the last decades.

A block of stamps from Palau.

Litho: W. Schmettkamp

Appendix

Other Symbiotic Clam Species

Since Brock discovered in 1888 the symbiosis between *Tridacnidae* and algae, it is known that these relationships exist between mussels and micro-organisms. However, the view that the giant clams are the only representatives of this style of living has proved to be wrong. In the meantime a number of subfamilies are known which live with unicellular organisms - in a variety of different ways - and which have adapted to their symbionts physiologically as well as anatomically. This way of living together is a special form of "symbiosis" (a way of living together where both partners benefit), called "trophobiosis" as the two partners benefit from mutual feeding.

The "trophobiontic" mussels can be divided into two groups: the symbiosis with unicellular algae and the symbiosis with bacteria. The first are represented by the *Tridacnidae* and the *Fraginae,* where the algae transfer sunlight into chemical energy through the process of photosynthesis (photoautotrophy).

The second form of symbiosis comprises many more families. While just less than 20 different species of clams live in symbiosis with algae, the group of clams that live in symbiosis with bacteria comprises almost 60 species. They belong to the following families: *Lucinidae, Fimbriidae, Solemyidae, Mytilidae, Vesicomyidae, Thyasiridae* and *Mactridae*. The symbiotic bacteria of these mussels get their energy from sulfids and methylated compounds and are therefore completely independent from light (chemoautotrophy).

Symbiosis with Algae: The Fraginae

The family *Fraginae* comprises many species. In the Philippine waters alone a considerable number are found: *Corculum cardissa, C. impressum, C. laevigatum, C. monstrosum, C. kirai, C. (Fragum) unedo, C. (Fragum) hemicardium, C. humanum, C. inversum* and *C. asebae*. Most of the species are closely related, which is demonstrated also by the similarity of their shells and their physiology. The very special shape of the shells and the way of living of these mussels, which is very different from the tridacnids, will be explained in the following taking *Corculum cardissa* as an example. This species was studied in the fifties by the Japanese biologist Kawaguti (Kawaguti, 1950, 1968) and later by the German biologist Janssen (Janssen, 1988, 1989).

In contrast to the tridacnids where the symbiotic algae live in the syphonal mantle which overlaps the shell, the *Fraginae* enclose the algae within their shells. This is a benefit on the one hand, as the soft body parts are not exposed to predators. On the other hand, this means that the sunlight has to penetrate

The unusually shaped mussel Corculum cardissa.

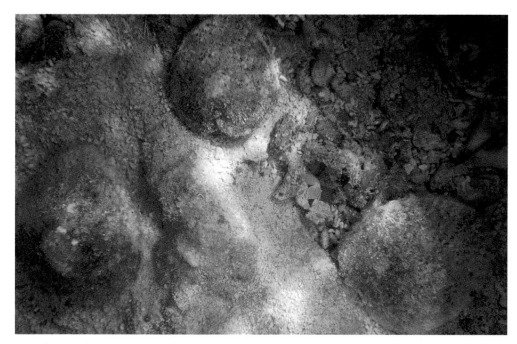

Corculum cardissa in its natural environment.   Photo: Dr. H.-H. Janssen

the shell to reach the algae. This problem has been solved by the fragines in the most interesting manner. The back side of the two shell halves is exposed to the light and has a large surface due to its flattened shape. These parts comprise numerous thin window-like structures. The sunlight penetrates these parts of the shell which resemble lenses and reaches the tissue with the symbiotic algae at the inside of the mussel.

This very interesting adaptation of the animals has, however, also its disadvantage which is a clear limit in growth. I believe that the symbiosis with unicellular algae which provide the mussel with an additional source of carbonate and which enables it to enhance its calcium synthesis, is the prerequisite for a faster growth rate. The form of symbiosis in tridacnids is suitable for a larger organisms, yet *C.*

cardissa and related species cannot exceed a certain maximal length, as they would have to thicken their shells to do so. This would of course decrease the light penetration of the shells which is in turn vital for the mussel. The *Fraginae* thus had to find a compromise between the thickest possible, most solid shell and the thinnest possible, most translucent shell. Increasing size was therefore a disadvantage for the *Fraginae,* while it is an advantage for the *Tridacnidae.*

The symbiotic algae of the *Fraginae* live in five different centers of the body. The highest density of algae is in the gills as well as in the tissue at bottom of the shell half. They are also found near the kidneys of the mussel and in lower concentrations below the upper shell in the vicinity of the lense-resembling windows. Similar to the tridacnids also

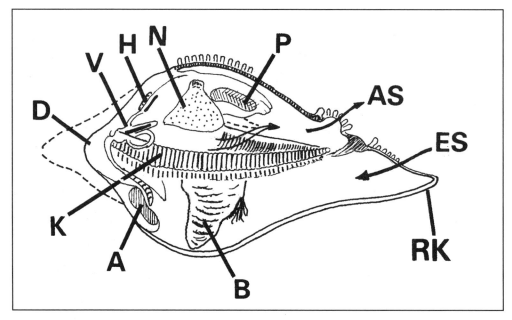

Physiology of Corculum cardissa seen from the side; one shell removed: A = anterior adductor muscle, B = byssal gland, D = dorsal side (back side), K = gills, V = digestive tract, H = heart, N = kidneys, P = posterior adductor muscle, AS = excurrent syphon, ES = incurrent syphon, RK = radial keel. As C. cardissa does not undergo the ontogenetic rotation of the Tridacnids (an anatomical transformation during metamorphosis), it belongs to the dimyaric mussels and thus possesses two adductor muscles.

Drawing by Melchior Buelo according to sketches by Dr. H.-H. Janssen

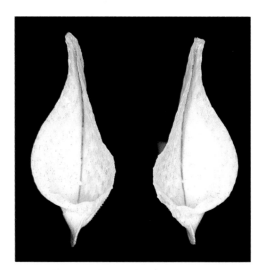

Upper shell side of Corculum cardissa. The translucent windows are visible in the middle of the picture.

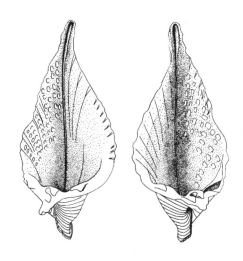

The lense resembling windows at the upper side of the shell are clearly visible on the left and right side of the shell in this drawing.

Drawing by Dr. H.-H. Janssen

240

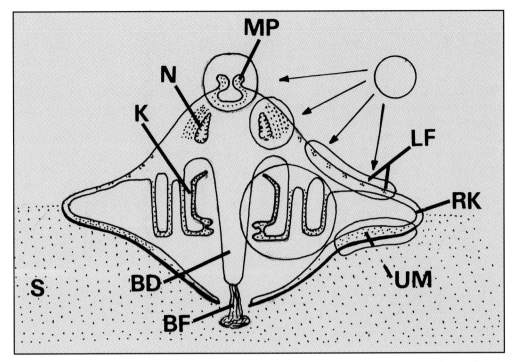

Horizontal section of Corculum cardissa (cross section through both shells). MP = mantle papilla, N = kidneys, K = gills, LF = lense window, RK = radial keel, UM = lower mantle, BD = byssal gland, BF = byssal filaments, S = sandy sediment.
Drawing by Melchior Buelo according to sketches by Dr. H.-H. Janssen

the fragines have developed a channel system for their symbionts, which is a network of tubes where the symbionts live. Janssen found in *Fraginae* at the same time as Norton in *Tridacnidae*, a three dimensional channel system stretching through the soft parts of the animal which is very different from the blood vessel system. The channels are translucent and have flagellated cells. Both systems develop from the digestion channel at an embryonic stage. The acquisition of symbiotic algae is also very similar in tridacnids and fragines: The symbionts are not transferred from one generation to the next; each juvenile has to acquire the algae from the sea water together with the nutrients by itself.

The phototrophic way of living of these *Fraginae* could mean that they are robust inhabitants of our aquaria. However, I have so far no experience with these mussels. Janssen reports of successful keeping of the species *C. cardissa.* Its survival in captivity is only curtailed, if its byssal apparatus has been injured during collecting (pers. comm.). I have searched intensively for these mussels in Cebu, but they are hard to find due to their excellent camouflage. The local fishermen, which know their places keep this secret very well. Also, a considerable part of the reefs in this country is unfortunately devastated due to dynamite fishing.

Fraginae and Tridacnidae, Families With a Common Ancestor?

According to Yonge (1953), the different shape of the shells in the *Tridacnidae* and the *Fraginae* indicate that they developed independently and he therefore believed that the acquisition of symbiotic algae also happened at least twice in evolution.

In contrast, Janssen, in a so far unpublished study, points out the parallels of the two families. He thinks that *Tridacnidae* and *Fraginae* have a common ancestor. It seems plausible that the two mollusc families with their marked similarities concerning the symbiont channel system, which seems to be unique in the animal kingdom, derive from a common ancestor.

Janssen also points out certain similarities between the two families concerning the shape of their shells. The species *Hippopus hippopus* has kind of a major fold in the middle of the vertical folds at the outside of the shell which protrudes particularly far and which divides the shell into a front and a back half. If the animal is moved around 90 degrees so that the incurrent syphon is on top, the similarity of this, now horizontal major fold with the keel-shaped rim fold of *C. cardissa*, becomes apparent. The other vertical folds which stabilize the shell of *H. hippopus* are indicated also in the shells of the *Fraginae*. They start from the whirl and run to the rim along the shell arch where they end with fine teeth which resemble quite a bit the teeth at the upper rim of the *H. hippopus* shells.

Because of these morphological and physiological similarities, an ancestor of the genus *Hippopus* could to my opinion be the link between the *Fraginae* and the *Tridacnidae*. The fact that the two *Hippopus* species lack an overhanging mantle point into the same direction. This would indicate that the overhanging mantle in the other

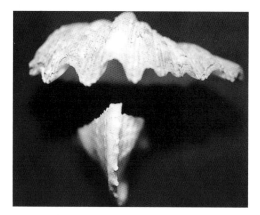

Confrontation of Hippopus hippopus and Corculum cardissa. The radial keel (Corculum cardissa) and the prominent major fold (Hippopus hippopus) can be seen very easily.

Underside of Corculum cardissa and Hippopus hippopus.

Tridacna species evolved later as an additional specialization.

It is, in my opinion much less likely that evolution went the other way round and that the genus *Hippopus* derived from the genus *Tridacna*. Obviously the family *Tridacnidae* derived from cardiine ancestors. At a certain point in evolution they began to cover their shell rims with the syphonal mantle. This adaptation must be advantageous as the *Tridacna* species follow this strategy until today. If the genus *Hippopus* is derived from the genus *Tridacna,* this would necessarily mean that the overhanging syphonal mantle has disappeared. Thus, a successful adaption to the environment would have disappeared. This also applies to the species *Tridacna tevoroa*, which does not have an overhanging syphonal mantle. The loss of this environmental adaption which has proven to be successful I regard as very unlikely.

In general the two *Hippopus* species appear much less well adapted to photobiosis. The number of their symbiotic algae is, for example, much lower than in the genus *Tridacna* (Yonge, 1980; Janssen, unpubl.) As a rule, the evolution of species goes into the direction of adaptive specializations as long as this is advantageous and not into the direction of reductive generalizations. Therefore I assume the genus *Hippopus* to be the eldest representative of the family *Tridacnidae.* The species *T. tevoroa* could have branched off before the tridacnids developed the overhanging syphonal mantle. More recently each species developed its species specific characters.

According to this hypothesis, it is likely that the *Fraginae* branched off before the development of the

The similarity between Fragum hemicardium, shown in this picture, and Hippopus hippopus is very obvious from a dorsal view into the byssal region.

ontogenetic rotation process undergone by all the members of the family *Tridacnidae.* The species specific adaptations of the *Fraginae* developed more recently. One of the *Fragum* species (*Fragum fragum*), for instance, developed an overhanging mantle. This development was, however, not homologous to the overhanging mantles of the *Tridacnidae,* but rather analogous, which means, that the same character exists in two species, but does not derive from a common ancestor. Another *Fragum* species (*Fragum unedo*) has formed the syphonal region into a plate containing the symbionts which can be protruded and exposed to the light. Altogether the *Fragum* species show signs of a longer axle and a radial keel while the backside shell is still undifferentiated (Janssen, 1988). On the one end of this development you would find the species *C. cardissa* which, like the *Tridacnidae* has specialized anatomically and physiologically to a high degree to live with its symbionts.

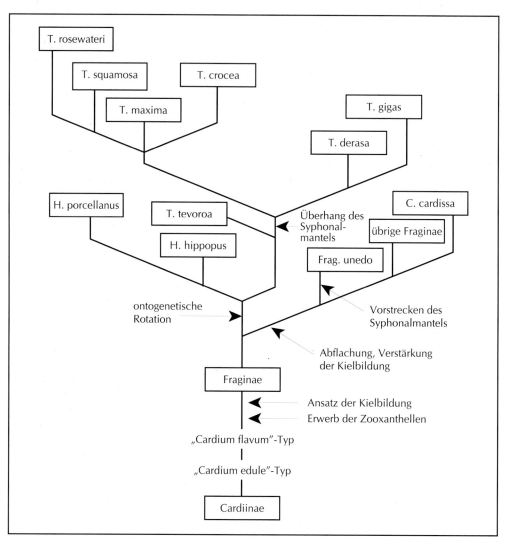

Model, how the Tridacnidae and the Fraginae could have evolved from a common, Cardium-like ancestor.

Symbiosis Between Mussels and Bacteria

Symbiosis between mussels and bacteria have only been discovered in the eighties. Our knowledge about these sulphidotrophic bacteria (which oxidate sulphur) is still very young and detailed research is in process. The facts about this way of metabolism, which was firstly discovered in deep sea organisms, has motivated various biologists to also examine animals in the shallow water. These symbiotic relationships for mutual nutrition are called "trophobiosis". In the meantime it has been shown that a number of mussel species follow this strategy of nutrition and there are continuously new species adding to this number.

Apart from the "sulphidotrophic" bacteria, also "methanotrophic" bacteria have been discovered, which gain their energy from chemical transformations of the gas methane (Childress et al., 1986; Fisher et al., 1987). I would think, that in this field we can expect a number of exciting and surprising new findings.

The mussels, which surprise many scientists with their way of nutrition today, are known since long. Many species of them are ever since a solid part of the diet of the local coastal populations. For example, the species *Codacia tigerina* is known amongst the Philippines as "tambayan", "limbuo" or "bakalan". The species *Anodontia edentula* and *Fimbria fimbriata* are offered under the names "imbaw" and "bug-atan" respectively at the local fish markets.

The new findings bring about the answers to many questions which were unsolved in science for a long time. The *Solemyidae* with their reduced digestive tract, for example, were known to biologists since decades, but their way of nutrition was unclear until recently. In 1961, people still believed, that in these mussels digestion takes place somewhere outside the digestive tract in the mantle cavity (Owen, 1961). In a more thorough investigation, 19 years later, this finding could not be confirmed (Reid, 1980), and another three years later, shortly after the "chemoautotrophic" sulphide bacteria had been discovered in the deep sea, the bacteria were discovered the the gill tissue (Felbeck, 1983). Motivated by these findings more searches were undertaken for similar symbiotic relationships and in the following years it was shown that the symbiosis between mussels and bacteria are much more widespread than originally anticipated.

In all investigated species the symbiotic bacteria always lived in special cells of the gill tissue (Janssen, 1992). It was assumed that these intra-cellularly living bacteria are transferred to the next generations by an infection of the eggs. The findings in this field are, however, still fragmentary. Some of these species (*Codacia tigerina, Fimbria fimbriata*) occur in sandy sediment rich patches of the coral reef, others, in contrast, inhabit the sandy and murky grounds of mangrove forests. In any case they need the organic substances rich waters, where the concentrations of sulphides are high. Thus, they inhabit niches were many other organisms cannot exist. They are able to detoxicate water bodies polluted by human influences and to transfer noxious waste into valuable protein rich substances. This ability could be helpful in the protein provision of countries of the Third World. Therefore I recommend to include, besides the *Tridacnidae,* also the mussels with their photoautotrophic and chemoautotrophic symbionts into captive propagation programmes as this could open future food sources.

References

Aquaristic Literature:

Hebbinghaus, R., 1994, Der Löbbecke-Kalkreaktor, in: Deutsche Aquarien und Terrarien Zeitschrift, Jahrg. 94, S. 517 - 525.

Wilkens, P., Niedere Tiere im tropischen Seewasseraquarium Bd. 1 und 2, Dähne Verlag, Ettlingen, Germany

Wilkens, P., Marine Invertebrates: Organ Pipe and Leather Corals, Gorgonians, Dähne Verlag, Ettlingen, Germany

Wilkens, P., Marine Invertebrates: Stony Corals, Mushroom and Colonial Anemones, Dähne Verlag, Ettlingen, Germany

Fosså, S. A. und Nilsen, A. J., Das Korallenriff-Aquarium, Bd. 1 - 3, Birgit Schmettkamp-Verlag, Bornheim, Germany

Scientific Literature:

ACIAR, 1986. Giant Clam project, 4th report. James Cook University of North Queensland. 13 p.

Alcala, A.C., Solis, E. P. and Alcazar, S. N., 1986, Spawning, larval rearing and early growth of Hippopus hippopus (Linn.) (Bivalvia: Tridacnidae), Silliman Journal, Vol. 33, No. 1-4, pp. 45-53

Alcazar, S. N., Solis, E. P. and Alcala, A. C., 1987. Serotonin-induced spawning and larval rearing of the China clam, Hippopus hippopus Rosewater (Bivalvia: Tridacnidae). Silliman Journal, 33, 65 - 72

Belda, C., et al. 1988. Effects of sediment and light intensities on the growth and survival of giant clams (Bivalvia: Tridacnidae). Paper presented to the Workshop on the Culture of Giant Clams in the Philippines, 15-17 March, 1988, Siliman University, Dumaguete City. 9 p.

Belda, C., et al. 1993 a, Modification of shell formation in the giant clam Tridacna gigas at elevated nutrient levels in sea water, Marine Biology 117, 251 - 257

Belda, C., et al. 1993 b, Nutrient limitation in the giant clam-zooxanthellae symbiosis: effcts of nutrient supplements on growth of the symbiotic partners, Marine Biology 117, 655 - 664

Bisker R. and Castagna M.,1985, The effects of various levels of air-supersaturated seawater on Mercenaria mercenaria (Linné), Mulinia lateralis (Say), and Mya arenaria (Linné), with reference to gas-bubble disease. Journal of Shellfish research, 5, 97 - 102

Blank, R., and Trench, R. K. 1985. Speciation in symbiotic dinoflagellates. Science, 229, 656-658.

Braley, R. D., 1985, Serotonin induced spawning in Giant clams (Bivalvia: Tridacnidae). Aquaculture, 47, 321 - 325

Braley, R. D., 1986, Reproduction and Recruitment of giant clams and some aspects of their larval and juvenile biology. Ph. D. Thesis, University of New South Wales, 297 p.

Braley, R. D. 1992, The Giant Clam: A Hatchery and Nursery Culture Manual, ACIAR Monogr. No 15, Canberra, Australia

Burrows, W. 1940. Notes on molluscs used as food by the Fijians. Fiji Society of Science and Industry, 2(1), 12-14.

Chalker, B. E. and Dunlap, W. C., 1984, Primary production and photoadaptation by corals on the Great Barrier Reef. In: Baker, J. T., Carter, R. M., Sammarco, P. W. and Stark, K. P. ed, Proceedings inaugural Great Barrier Reef Conference. Townsville, August 28 to September 2, 1983, JCU Press, 293 - 298

Chang, S.S., Prezelin, B.B., and Trench, R.K. 1983. Mechanisms of photoadaption in three strains of the symbiotic dinoflagellate Symbiodinium microadriaticum. Marine Biology, 76, 219-229.

Childress, J. J. et al. 1986, A methanotrophic marine molluscan (Bivalvia: Mytilidae) symbiosis: Mussels fueled by gas. Science 233: 1306 - 1308

Crawford, C. M. et al., 1986, Spawning induction and larval and juvenile rearing of the giant clam, Tridacna gigas, Aquaculture, 58, 281 - 295

Crawford, C. M. et al., 1987. The mariculture of giant clams. Interdisciplinary Science Reviews,12,333-340.

Crawford, C. M. et al., 1988. Growth and survival during the ocean-nursey rearing of giant clams, Tridacna gigas. 1. Assessment of four culture methods. Aquaculture, 68, 103-113.

Deane, E.M., and O'Brien, R.W. 1980. Composition of haemolymph of Tridacna maxima (Mollusca:Bivalvia). Comparative Biochemistry and Physiology, 66, 339-341

Digby,P.S.B. 1968. The mechanism of calcification in the molluscan shell. Symposium of the Zoological Society, London, 22, 93-107.

Eibl-Eibesfeldt, Irenäus, 1971, Liebe und Haß, R. Piper-Verlag, München, 4. Aufl., S. 35

Estacion, J., Solis, E. and Lourdes, F., 1986, A preliminary study of the effect of feeding on the growth of Tridacna maxima (Röding) (Bivalvia: Tridacnidae), Silliman Journal Vol.33, No. 1-4, pp. 111-116

Estacion, J. C. 1988. Ocean nursery phase. Paper presented to the Workshop on the Culture of Giant Clams in the Philippines, 15-17 March,1988, Silliman University, Dumaguete City. 11p.

Felbeck, H. 1983, Sulfide oxydation and carbone fixation by the gutless clam Solemya reidi (Bivalvia: Protobranchia)

Fisher, C. R. et al. 1985. Photosynthesis and respiration in Tridacna gigas as a function of irradiance and size. Biological Bulletin, 169, 230-245.
Fisher, C. R. et al. 1987, The importance of methane and thiosulfate in the metabolism of the bacterial symbionts of two deep sea mussels. Mar. Biol. 96:59 - 71

Fitt, W K. and Trench, R. K., 1980. Uptake of zooxanthellae by veliger and juvenile stages of Tridacna squamosa. American Zoologists, 20. 777.

Fitt, W. K. and Trench, R. K., 1981, Spawning, developement and aquisition of zooxanthellae by Tridacna squamosa (Mollusca: Bivalvia), Biol. Bull. 161: 213 - 235

Fitt, W. K. et al. 1984. Larval biology of tridacnid clams. Aquaculture, 39, 181-195, 1986. Contribution of the symbiotic dino-flagellate Symbiodinium microadriaticum to the nutrition, growth, and survival of larval and juvenile tridacnid clams. Aquaculture, 55, 5-22.

Fitt, W. K., 1988, Increasing growth rates and survival of giant clams in mariculture. J. Shellfish Res., 7:547,

Fretter, V., and Graham, A. 1949. The structure and mode of life of the Pyramidellidae, parasitic opisthobranchs. Journal of the Marine Biology Association U.K., 28, 493-532.

Gomez, E. D. und Alcala, A. C., 1988, Giant clams in the Philippines, in: Copland, J. W. and Lucas, J. S, (ed.), Giant Clams in Asia and the Pacific, Canberra, Australia, ACIAR Monograph Nr. 9

Govan, H., Nichols, P.V., Tafera, H., 1988, Giant Clam Resource Investigations in Solomon Islands, in: Copland, J. W. and Lucas, J. S, (ed.), Giant Clams in Asia and the Pacific, Canberra, Australia, ACIAR Monograph Nr. 9

Hamner, W.M., 1978. Intraspecific competition in Tridacna crocea, a burrowing bivalve. Oecologia, 34, 267 - 281

Heslinga, G. A. und Fitt, W. K., 1987, The domestication of tridacnid clams, Bioscience, 37, 332 - 339

Huguenin, J.E and Colt, J. 1989. Design and operating guide for aquaculture seawater systems. Developements in Aquaculture and Fisheries Science, 20. Elsevier, Amsterdam, 264 p.

Janssen, H.-H., 1988, Corculum cardissa - symbionteninduzierte Co-Evolution bei Herzmuscheln (Cardiaceae: Fraginae). Verh. Dtsch. Zool. Ges. 81:186

Janssen, H.-H., 1989, Morphology and ultrastructure of the heart shell, Corculum Cardissa (Bivalvia: Cardiacea: Fraginae), a consequence of adaptation to endosymbiotic zooxanthellae, 10th internat. Malacol. Congr., Tübingen, Aug. 27th -Sept. 2nd, 1989

Janssen, H.-H., 1992, Philippine Bivalves and Microorganisms, Past Research, Present Progress and a Perspective for Aquaculture. The Philippine Scientist 29 (1992): 5 -32

Janssen, H.-H., unveröffentl., Zooxanthellae in Cardiacea (Bivalvia): Comparative Review and Implications for Evolution

Jameson, S. C., 1976, Early life history of the giant clams Tridacna crocea (Lamarck), and Tridacna maxima (Röding), and Hippopus hippopus (Linnaeus), Pacific Science, 30, 219 - 233

Kawaguti, S., 1950, Observations on the heart shell, Corculum cardissa (L.), and its associated Zooxanthellae. Pacific Science 4, 43 - 49

Kawaguti, S., 1968 Electron microscopy on zooxanthellae in the mantle and gill of the heart shell. Biol. J. Okayama Univ. 14 (1-2), 1 - 11

Kawaguti, S., 1983, The third record of Association between Bivalve Molluscs and Zooxanthellae. Proc. Japan Acad. 59, Ser. B, 17 - 20

Kinsey, D. W., and Davies, P. J., 1979, Effects of elevated nitrogen and phosphorus on coral-reef growth. Lomnol. Oceanogr. 24: 935 - 940

Kinsey, D. W., 1991, Water quality and its effect on Reef ecology. In: Yellowlees, D. (ed.), Land use patterns and nutrient loading of the Great Barrier Reef Region. James Cook University Press, Australia, p. 192 - 196

Knop, D., 1994 a, Giant Clam Export Philippines to Germany, Survival Rates, in: Clamlines, Newsletter of the Giant Clam Research Group, No. 14, ICLARM-CAC, Solomon Islands

Knop, D., 1994 b, Whitespot Disease, Infectious Disease in Aquarium Clams, in: Clamlines, Newsletter of the Giant Clam Research Group, No. 14, ICLARM CAC, Solomon Islands

Knop, D., 1994 c, Giant Clam Stocking Tanks for Restaurants, in: Clamlines, Newsletter of the Giant Clam Research Group, No. 14, ICLARM-CAC, Solomon Islands

La Barbera, M., 1975, Larvae and larval developement of the giant clams Tridacna maxima and Tridacna squamosa Bivalvia: Tridacnidae). Malacologia, 15 (1), 69 - 79

Lee, P. S., 1990, Aspects of the Biology of Metamorphosis in Tridacnid Clams, with Special Reference to Hippopus hippopus. M.S. Thesis, James Cook University, Townsville, Queensland.

Lewis, A. D. and Ledua, E., 1985, A Possible New Species of Tridacna (Tridacnidae: Mollusca) from Fiji. In: Copland, J. W. and Lucas, J. S, (ed.), Giant Clams in Asia and the Pacific, Canberra, Australia, ACIAR Monograph Nr. 9

Lewis, A. D., Adams, T. J. H. and Ledua, E., 1988, Fiji's Giant Clam Stocks, A Review of their Distribution, Abundance, Exploitation and Management. In: Copland, J. W. and Lucas, J. S, (ed.), Giant Clams in Asia and the Pacific, Canberra, Australia, ACIAR Monograph Nr. 9

Littlewood, D.T. J. and Marsbe, L.A., 1990. Predation on cultivated oysters, Crassostrea rhizophorae (Guilding), by the polyclad turbellarian flatworm, Stylochus frontalis Verrill. Aquaculture, 88, 145 - 150

Luckner, G.1983, Diseases of Mollusca: Bivalvia. In: Kinne, O., ed., Diseases of marine animals, vol. II. Hamburg, Biologische Anstalt, 477 - 961

MacDonald, J. D. 1854. On the anatomy of Tridacna. Proceedings of the Royal Society London, 7, p.589-590.

Malouf, R. et al., 1972, Occurence of gas-bubble disease in three species of bivalve molluscs. Journal of the Fisheries Research Board Canada, 29, 588 - 589

Mansour, K., 1946, Communication between the dorsal edge of the mantle and the stomach of Tridacna. Nature (Lond.), 157:844

Mingoa, S. S., 1988, Photoadaptation in Juvenile Tridacna gigas, in: Copland, J. W. and Lucas, J. S, (ed.), Giant Clams in Asia and the Pacific, Canberra, Australia, ACIAR Monograph Nr. 9, 145 - 150

Mingoa-Licuanan, S. S., 1993, Oxygen consumption and ammonia excretion in juvenile Tridacna gigas (Linné, 1758): effects of emersion. Mar. Biol. Ecol., 171: 119 - 137

Murakoshi, M. 1986. Farming of the boring clam, Tridacna crocea (Lamarck) Galaxea, 5, 239-254.

Norton J. H., Sheperd M.A., Long H.M. and Fitt W.K., 1992, The Zooxanthellal Tubular System in the Giant Clam, Biol. Bull. 183, 503 - 506

Norton, J.H., Sheperd, M. A., Abdon-Naguit, M. R. and Lindsay, S., 1993, Mortalities in the giant clam Hippopus hippopus Associated with Rikkettsiales-like Organisms. Journal of Invertebrate Pathology 62, 207-209

Norton, J. H., Perkins, F. P. and Ledua, E. 1993, Marteilia-like Infection in a Giant Clam, Tridacna maxima, in Fiji, Journal of Invertebrate Pathology 61: 328-330

Norton, J. H. et al. 1993, Intracellular Bacteria Associated with Winter Mortality in Juvenile Giant Clams, Tridacna gigas, Journal of Invertebrate Pathology 62: 204-206

Owen, G. 1961, A note on the habits and nutrition of Solemya parkinsoni (Protobranchia: Bivalia). Q. J. Microscop. Sci.102: 15 - 21

Pasaribu, B. P., 1988, Status of Giant Clams in Indonesia, in: Copland, J. W. and Lucas, J. S, (ed.), Giant Clams in Asia and the Pacific, Canberra, Australia, ACIAR Monograph Nr. 9

Perkins, F. O., 1985, Range and host extensions for the molluscan bivalve pathogens, Perkinsus spp. Abstract, VII International Congress of Protozoology, 1985, Nairoby, Kenia, p. 81

Rasmussen, C., 1989, The use of strontiom as an indicator of anthropogenically altered environmental parameters, Proc. 6th int. Coral Reef Symp. 2: 325 - 339 [Choat, J. H., et al. (eds) Sixth International Coral Reef Symposium Executive Committee, Townsville]

Reid, R. G. B. 1980, Aspects of the biology of a gutless species of Solemya (Bivalvia: Protobranchia). Can. J. Zool. 58: 386 - 393

Reid, R. G. B., and Brand, D.G. 1986. Sulfide oxidising symbiosis in lucinaceans: implications for bivalve evolution. Veliger, 29, 3-24.

Reid, R. G. B., and Slack-Smith, S. 1988. The Lucinacea. In Fauna of Australia. 5. Australian Bureau of Flora and Fauna, Canberra. (In Press.) Richard, G., 1985, Richness of the great sessile bivalves in Takapoto lagoon. Proceedings of the Fifth International Coral Reef Congress, 1, 368 - 371

Rosewater, J. 1965. The family Tridacnidae in the Indo-Pacific. Indo-Pacific Mollusca,1, 347-396. 1982. A new species of Hippopus (Bivalvia:Tridacnidae). Nautilus, 96, 3-6.

Sirenko, B. I. and Scarlato, O. A. 1991, Tridacna rosewateri, A new species of giant clam from Indian Ocean. La Conchiglia No. 261: 4-9

Solis, E. P. and Heslinga, G. A., 1989, Effect of Desiccation on Tridacna derasa Seed: Pure Oxygen Improves Survivals During Transportation. Aquaculture 76: 169-172

Stephenson, A., 1934, The breeding of reef animals, Part II, Invertebrates other than corals, Scientific Report Great Barriere Reef Expedition, 1928 - 29, 3, 247 - 272

Syárani, L., 1987, The exploration of giant clam fossils on the fringing reef areas of Karimun Jawa Islands. Biotrop Special Publication 29, 59 - 64

Taniera, T., 1988, The Status of Giant Clams in Kiribati, in: Copland, J. W. and Lucas, J. S, (ed.), Giant Clams in Asia and the Pacific, Canberra, Australia, ACIAR Monograph Nr. 9

Thomas, P. A., 1979, Boring sponges of destructive to economically important molluscan beds and coral reefs in indian seas. Indian Journal of Fisheries, 26 (1-2), 163 - 200

Trench, R. K. et al. 1981. Observations on the symbiosis with zooxanthellae among the Tridacnidae (Mollusca, Bivalvia). Biological Bulletin, 161, 180-198.

Trinidad-Roa, M. J., and Alialy, E. O. 1988. Mariculture of giant clams in Bolinao, Pangasinan. Paper pressented at the National Giant Clams Workshop, Silliman University, Dumaguete, Philippines, 15-17 March 1988.

UPMSI. 1987. Report of the University of the Philippines Marine Sciences Institute, Annex to the Thirty-Six Month Giant Clam Report. July 1986-June 1987.

Usher, G. F., 1984, Coral reef invertebrates in Indonesia. Their exploitaition and conservation needs. Rep. IUCN/WWF Project 1688, Bogor IV, 100 p.

Wada, S., 1954, Spawning of tridacnid clams. Japan Journal of Zoology, 11, 273 - 285

Yonge, C. M., 1936, Modes of life, feeding, digestion and symbiosiswith zooxanthellae in the Tridacnidae. Scientific Reports on the Great Barrier Reef Expedition 1928 - 29, Vol. 1, 283 - 321.

Yonge, C.M., 1953 a, Mantle chambers and water circulation in the Tridacnidae (Mollusca), Proc. Zool. Soc. (Lond.), 123: 551 - 561
Yonge, C. M., 1953 b, The monomyarian condition in the lamellibranchia, Trans. R. Soc. Edinb. 62: 443 - 478

Yonge, C. M., 1975, Giant Clams, Scientific American, 23, 96 - 105

Yonge, C. M., 1980, Functional Morphology and evolution in the Tridacnidae (Mollusca: Bivalvia: Cardiacea). Records Australian Museum 33 (17): 735 - 777, Figs. 1 - 29

Index